# The Internal Environment of Dwellings

R. M. E. DIAMANT

# The Internal Environment of Dwellings

HUTCHINSON EDUCATIONAL

HUTCHINSON EDUCATIONAL LTD
*178–202 Great Portland Street, London W1*

London Melbourne Sydney Auckland
Wellington Johannesburg Cape Town
and agencies throughout the world

*First published 1971*





*This book has been set in Imprint type, printed in Great Britain on smooth wove paper by Anchor Press, and bound by Wm. Brendon, both of Tiptree, Essex*

ISBN 0 09 107300 6

Acknowledgements

My thanks are due to the following firms for permission to reproduce illustrations: Fibreglass Ltd. for Figure 9; Pilkington Brothers Ltd. for Figure 10; Deighton Wilkes & Company Ltd. for Figure 13; Onazote Ltd. for Figure 15; Courtaulds Ltd. for Figure 18; Thermalon Ltd. for Figure 23; Alcan Ltd. for Figure 24; and Futurum AB, Sweden, for Figure 25.

# Contents

# Preface

When designing buildings, rigorous calculations are always carried out with regard to the foundations, wind loadings, floor and roof loadings and so on. Yet seldom, if ever, is there much sign that equal effort has been expended to ensure that the dwelling is properly insulated and free from condensation.

This book has been written to teach those concerned with building how one can obtain a good internal environment. Although a quantitative approach is being used, the actual mathematics needed is fourth form school standard.

It is intended that this book becomes an essential text for technical college and technical institute students of all branches of building, and also for students of architecture, town planning and so on.

In addition, the book is considered suitable for practising builders, architects, heating engineers, civil engineers and so on, as well as for their customers. This means local authorities, estate agents and developers, as well as members of the general public who wish to ensure that the house which is being built for them is really going to be suitable to live in.

The British building industry is now going metric, and for this reason this book is using SI metric units only.

A significant portion of the text first appeared as a series of papers in the journal *Building* between September 1969 and January 1970.

R. M. E. DIAMANT
*University of Salford*
*1970*

# I Introduction

Several specialists play their part when buildings are being constructed. The town planner designs the layout so that adequate room is left for roads and open spaces, the architect is responsible for the general aesthetic appearance and the internal layout, while the structural engineer is mainly concerned with the engineering design of the building.

But the internal environment of the building, by which one includes the general temperature level, freedom from moisture penetration and condensation, as well as protection from internal and external noise, appears to be no one's responsibility. Guesswork, doing what has been done before, and slavish compliance with the letter of outdated building regulations appear to be the order of the day.

The results of all this we know only too well. Some 50% of all recently erected dwellings suffer badly from condensation, which is of such a magnitude in some cases that homes have had to be evacuated altogether. Luxurious prestige office blocks, on which unlimited sums of money have been lavished, are impossible to work in during the summer.

All these troubles can be avoided quite easily, and it is often amazing how little vital improvement in internal environmental conditions costs. But it is a field which cannot be dealt with qualitatively. One must apply modern scientific thought to the problems and come up with a numerical answer. After all, the structural engineer does not just place his hands like the angler describing 'the one that got away' and assume that this should be the cross-sectional area of the stanchions. He uses mathematical methods to calculate the exact thickness of the stanchions.

## Use of metric units[1]

All the values that are being used are completely different from those readers are likely to be accustomed to. The reason is that this book uses SI metrication units only. After all, the building industry will be going metric very shortly, and by 1975 the entire British industry will be working entirely in metric units.

Apart from the fact that all the calculations are far easier to carry out in metric units than in imperial units, it is vital that architects and builders start to think metric. This means that the sooner we start to remember factors such as *U*-values in terms of watt/metre$^2$ K instead of Btu/hour ft$^2$ °F, the better for all concerned. Students who started university and technical college courses in 1968 or 1969 are all trained in the new units only, and established architects, building and civil engineers can certainly not afford to be unfamiliar with the new international units.

## The main topics

This book deals with the following main topics:

(1) The heat balance of a building, including methods of heating, air conditioning, comfort requirements, insulation and associated topics

(2) The prevention of moisture penetration of the walls of a building, condensation against walls or within insulation spaces, the provision of ventilation and the effect of ventilation and insulation upon the liability of a building to be affected by condensation

# 2 Heat Balance

## Heat gain

To study heat balance we must sum up the main sources of heat and the main sources of heat losses. The three main sources of heat are:

(1) heat from electric lights, electric motors and other electric appliances
(2) heat given off by the occupants of a dwelling
(3) heat radiated through the windows

*Heat from electrical equipment*

When an electric motor is running, electrical energy is being converted into mechanical energy. But this mechanical energy is later being converted into heat energy. The first law of thermodynamics says that energy cannot be created nor destroyed, and therefore one can find out just how much heat is being supplied to a dwelling simply by adding up all the wattages of any electric motors used. Under SI all power values are measured in watts anyway, so there is no awkward converting necessary. All one has to do is to add, remembering, of course, that kilowatt means 1 000 watt. If the motors are still being quoted in terms of horsepower, one must convert 1 horsepower =746 watt.

In the case of electric lights, electric energy is being converted into heat energy and light energy. As is well known, fluorescent lighting produces a far higher proportion of light energy than do the normal filament lamps. But as the light impinges upon walls in the room, it also is changed into heat. If we imagine a room properly curtained at night, virtually all the light energy is eventually converted into heat energy. Under such circumstances, it is again correct to add up simply the wattages of the electric lights used and count them as a heat gain. When windows are left uncurtained, a proportion of the electric light escapes and, therefore, the heat gain will be lower than the wattage of the lamp. But in most cases such losses will be small and can probably be ignored. Even refrigerators can be considered to be sources of heat. Again one simply takes the wattage as the numerical value.

A simple way to determine the heat gain of a dwelling due to electrical appliances is to take electricity meter readings at different times of the day. For example, let us consider that the electricity meter stands at 405693·2 units or kilowatt hours at 12 noon one day and at 405712·5 units at 12 noon the following day. This means that 19·3 kilowatt hours of electricity have been consumed in 24 hours. The heat contribution made by electric appliances averaged out over 24 hours is thus simply:

$$\frac{19\,300}{24}\text{ watt} = 805\text{ watt}$$

If a study is to be made of how the heat contribution of electric appliances varies over the day, it is an easy matter to take electricity meter readings over shorter periods, say two hours, and again divide the number of kilowatt hours by the number of hours during which the power was consumed. In the near future, electricity may not be charged in the form of kilowatt hours but in the form of megajoules (1 000 000 joules). When this happens all that will be necessary to is divide the number of joules by the number of seconds during which the power was consumed, i.e. number of megajoules multiplied by 1 000 000 and divided by number of seconds.

We have to remember that a watt is in fact:

$$\frac{\text{joule}}{\text{second}}$$

This is very important.

*Heat given off by occupants*

Nearly all the food that people eat is converted by natural processes into body heat and into mechanical energy. This mechanical energy enables us to lift articles, move around, etc. But in the final analysis even this mechanical energy is once more converted into heat. Some of the calorific value of the food is converted into stored fat, but this is in general not a large percentage even if one is inclined to put on weight. It can be ignored in our calculations.

Using basic physiological data,[2] we can evaluate the contribution to space heating made by a human being as follows (over a 24-hour period):

| *Men* (average weight 70 kg)<br>(age range) | watt |
|---|---|
| 18–35 | 145 |
| 35–55 | 130 |
| 55–75 | 110 |

| *Women* (average weight 58 kg)<br>(age range) | watt |
|---|---|
| 18–35 | 105 |
| 35–55 | 95 |
| 55–75 | 80 |

| *Children* Boys (age range) | watt | Girls (age range) | watt |
|---|---|---|---|
| 1– 3 | 65 | 1– 3 | 65 |
| 3– 6 | 80 | 3– 6 | 80 |
| 6– 9 | 105 | 6– 9 | 105 |
| 9–12 | 120 | 9–12 | 110 |
| 12–15 | 150 | 12–15 | 125 |
| 15–18 | 170 | 15–18 | 115 |

These are average figures over the whole day. To obtain a figure for specific activities, one assumes the following values:

when asleep: deduct 20–30 watt
when sitting: take given value
when undergoing gentle activity: add 10–20 watt
when undergoing strenuous activity: add 30–40 watt

As an example, the heat given off by occupants in the case of a teenage dance club is the following, assuming that 100 males and 100 females between the ages of 15 and 18 are present, and that they are engaged in strenuous activity.

$100 \times (170+40)$ watt $= 21\,000$ watt from the boys
$100 \times (115+40)$ watt $= 15\,500$ watt from the girls

The total amount of heat produced is thus equal to 36 500 watt or 36·5 kilowatt. Heating would hardly be required even in midwinter. In fact, the basic requirement would be good air conditioning.

In contrast, let us consider the sitting room of a home for retired gentlewomen. Assuming that it caters for 50 occupants, the heat from their bodies will only be:

$$50 \times 80 \text{ watt} = 4000 \text{ watt} = 4 \text{ kilowatt}$$

*Heat gains through windows*

Solar radiation through windows makes a very important contribution to the heating of a building. It is quite surprising how much heat can be gained in this way, even on the coldest of days in midwinter. At its best, solar radiation inwards through the windows can markedly reduce fuel bills. It is possible, by suitable design of the building, using a very high standard of insulation and making use of solar heat and waste heat from the various electrical appliances in the room, virtually to eliminate any artificial forms of heating. This was done some years ago with a school in Wallasey, Cheshire, where no other forms of heating whatever were needed. In most cases, however, such an ideal is not readily achieved.

At its worst solar radiation through windows may make premises quite impossible to live and work in. Many of the newer types of office blocks with very large window areas suffer from these troubles.

The amount of heat which is gained through the windows is markedly dependent upon the following factors:

(a) *The compass direction into which the windows face.* In winter, the maximum amount of heat gain is obtained through windows facing south. In spring and autumn, the windows facing south-east get the greatest heat gain in the morning, and the windows facing south-west get the greatest heat gain in the afternoon. In summer the rooms facing east tend to be the hottest in the morning and those facing west the hottest in the afternoon. Surprisingly enough, rooms facing due south do not get over-hot in summer. They register their maximum heat gain in March and September.

(b) *The time of day.* In almost all cases the bulk of the heat passing through the window goes through in a few hours of the day only. Again, the season makes a difference. The maximum radiation inwards takes place in winter around midday. But in spring and autumn it is around 9 a.m. in the morning and 3 p.m. in the afternoon, while in summer the maximum amount of radiation may take place as early as 7 a.m. in the morning and 6 p.m. in the evening. Naturally the windows most affected will be those facing east in the morning and those facing west at night.

(c) *The clarity of the air.* Rain, fog or cloudiness markedly reduce the degree of solar radiation through the windows, and naturally at night the degree of solar radiation will be zero. On a cloudy day the radiation through the windows will approach that usually associated with windows facing north. Allowances must also be made for shading of the window. If part of the window is not subjected to direct sunlight, the value for north-facing windows should be taken for this section, whatever the compass direction. This will vary with the time of day and allowances for the length of the shadow must be made.

(d) *The nature of the windows.* The values in the tables apply to normal single glazing. For different types of windows these values must be multiplied by the following factors:[3]
Double glazing: multiply by 0·9
Double glazing + light plastic Venetian blind: multiply by 0·33
Single glazing + light plastic Venetian blind, light net curtain, etc.: multiply by 0·5
Heat absorbing glass: multiply by 0·6
Reflecting glass: multiply by 0·3

(e) *The geographic latitude of the building.* The figures quoted have been adapted from the 1968 report by I. Höglund and D. G. Stephenson[4] on solar radiation in Canada, dealing with work done for the Canadian National Research Council. The figures apply to a latitude of 56°N., which corresponds in this country to a line running from Glasgow to Edinburgh. Readers resident in this geographic area can take the figures as they stand. Readers living in other areas of the country must apply the correction factors given in Table 2.

TABLE 1. Solar radiation through windows in watt/metre², latitude 56°N.[4]

| *Compass direction of window* | *North* | *North-East* | *East* | *South-East* | *South* | *South-West* | *West* | *North-West* |
|---|---|---|---|---|---|---|---|---|
| *Month* | | | | | | | | |
| January | 5 | 5 | 30 | 75 | 75 | 30 | 5 | 5 |
| February | 10 | 10 | 60 | 120 | 105 | 35 | 10 | 10 |
| March | 15 | 35 | 105 | 155 | 110 | 35 | 15 | 15 |
| April | 20 | 65 | 135 | 160 | 95 | 35 | 20 | 20 |
| May | 35 | 100 | 165 | 160 | 80 | 35 | 20 | 20 |
| June | 40 | 110 | 170 | 145 | 75 | 35 | 25 | 25 |
| July | 30 | 100 | 160 | 155 | 80 | 35 | 20 | 20 |
| August | 20 | 60 | 130 | 155 | 95 | 35 | 20 | 20 |
| September | 15 | 30 | 95 | 145 | 105 | 35 | 15 | 15 |
| October | 10 | 10 | 55 | 110 | 100 | 35 | 10 | 10 |
| November | 5 | 5 | 30 | 70 | 70 | 35 | 5 | 5 |
| December | 3 | 3 | 15 | 50 | 55 | 25 | 3 | 3 |

Table 1 gives the amount of solar radiation flowing through windows expressed in watt/metre². The conditions which apply are those relating to latitude 56°N (Glasgow–Edinburgh line) and clear sky. Although the values give the average over 24 hours it must be realised that in fact nearly all this heat flows in during about seven hours in winter and perhaps 16 hours in summer only, and that hourly fluctuations even during the day are enormous. In most cases the maximum degree of radiation through the windows is between five and eight times as high as the average flow rate given in the table. In the case of the windows facing south-west the maximum degree of radiation inwards can be as much as 15 times as high as the daily average quoted, which means that nearly all the radiation is contributed over a period of two or three hours only, usually at noon.

Radiation is proportional to the difference between the fourth power of the absolute temperature of the radiating surface (the sun) and the fourth power of the absolute temperature of the receiving surface (the window).

$$Q = e\ s\ (T^4_{sun} - T^4_{window})$$

where *e* is the emissivity factor of the glass (about 0·6), *s* is the

Stefan–Boltzmann factor ($5{\cdot}68 \times 10^{-8}$ W/m$^2$ K$^4$) and $Q$ is the heat received in W/m$^2$

In the morning the window surface is cold and therefore the amount of radiation received is greater than in the afternoon, when the window has already warmed up. For this reason windows facing east seem to get more heat radiation than windows that face west. This was also borne out experimentally in all the results obtained.

For places other than towns close to the Glasgow–Edinburgh axis the following corrections have to be made. As can be seen they vary considerably with the time of the year.

TABLE 2. Correction factors as a percentage

| *Month* | Jan. Feb. | Mar. April | May June | July Aug. | Sept. Oct. | Nov. Dec. |
|---|---|---|---|---|---|---|
| *Latitude* | | | | | | |
| 52°N (Southern England) | +36 | +16 | −5 | −7 | +8 | +37 |
| 53°N (Midlands) | +27 | +12 | −4 | −5 | +6 | +28 |
| 54°N (Lancashire and Yorkshire) | +18 | +08 | −3 | −4 | +4 | +18 |
| 55°N (Newcastle–Carlisle belt) | + 9 | + 4 | −1 | −2 | +2 | + 9 |
| 56°N (Edinburgh–Glasgow belt) | 0 | 0 | 0 | 0 | 0 | 0 |
| 57°N (Aberdeen) | − 9 | − 4 | +1 | +1 | −2 | − 9 |

WORKED EXAMPLE

Let us determine the average solar radiation through an office window, facing south-east and measuring 3 m × 2 m, in a block situated in Birmingham during a 24-hour period in June. Looking at Table 1 we obtain a value of 145 watt/metre$^2$ for a window facing south-east during the month of June. The window area is 2 m × 3 m = 6 m$^2$ so that the value for latitude 56°N is equal to:

$$6 \times 145 \text{ watt} = 870 \text{ watt}$$

But the correction factor for Birmingham in June is −4%, so that we have to multiply this value by 0·96 to get the correct answer.

$$870 \times 0{\cdot}96 = 845 \text{ watt}$$

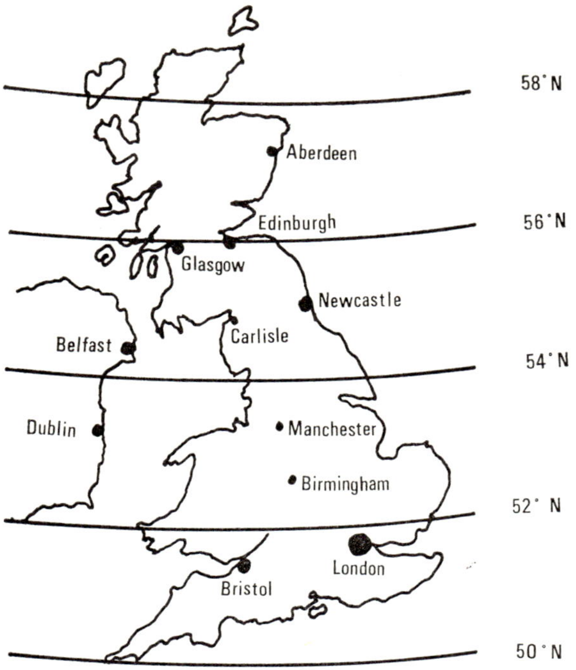


FIG. 1 Comparison of latitudes of various places in the British Isles

Over a 24-hour period it is thus necessary to get rid of $845 \times 3600$ (seconds/hour) $\times 24$ joules of heat or approximately 73 megajoules. If the windows in question are fitted with light plastic Venetian blinds, the amount of solar radiation is reduced to half, i.e. about 420 watt, while if reflecting glass is fitted, the radiation inward is equal to one-third or 280 watt. The amount of heat which must be got rid of during 24 hours to maintain a suitable heat balance would then equal 36 and 24 megajoules respectively, plus the heat produced by occupants and electric appliances.

*Maximum heat flow during day*

To determine the capacity of air conditioning equipment it is necessary to know the maximum rate of heat flow through the windows, the thermal storage effect of the building and the compass direction from which the heat enters. The period of the day when solar radiation is at maximum is also important. These data are given in Table 3.

TABLE 3. Maximum amount of heat radiation through windows at different times of the year. Latitude 56°N, but figures can be used for other places in the UK without introducing excessive errors

| *Month* | *Maximum heat flow (watt/metre²)* | *Compass direction* | *Time of day* |
|---|---|---|---|
| January | 705 | south | noon |
| February | 795 | south | noon |
| March | 770 | south-east, south, south-west | 10 a.m.–2 p.m. |
| April | 745 | south-east, south-west | 9 a.m. |
| May | 745 | east | 7 a.m. |
| June | 735 | east | 7 a.m. |
| | | west | 5 p.m. |
| July | 730 | east | 7 a.m. |
| | | west | 5 p.m. |
| August | 715 | south-east | 10 a.m. |
| | | south-west | 2 p.m. |
| September | 745 | south | noon |
| October | 755 | south | noon |
| November | 680 | south | noon |
| December | 620 | south | noon |

It can be seen that the maximum rate of solar radiation during the day varies hardly at all with the month of the year. In fact it is also independent of the latitude and therefore the values from Table 3 can be used throughout the British Isles.

The maximum heat flow through windows facing north fluctuates between 45 and 215 watt/metre² over the year.

*Horizontal glass areas (roof lights)*
For these the data given in Table 4 are applicable. They are once more given with respect to latitude 56°N (Glasgow–Edinburgh axis), and the average figures over the day as well as the maximum values must be loaded according to the factors given in Table 2 if other parts of the British Isles

are considered. The maximum degree of solar radiation is always at noon.

TABLE 4. Radiation through horizontal roof lights

| | *Average solar radiation (watt/metre²)* | *Maximum solar radiation (watt/metre²)* |
|---|---|---|
| January | 15 | 155 |
| February | 35 | 310 |
| March | 70 | 480 |
| April | 105 | 635 |
| May | 135 | 720 |
| June | 150 | 750 |
| July | 135 | 720 |
| August | 105 | 625 |
| September | 65 | 465 |
| October | 35 | 300 |
| November | 15 | 150 |
| December | 10 | 100 |

We have now examined the main sources of heat gain in a dwelling, and summed them up as:

(1) heat gain from electric equipment
(2) heat gain from occupants
(3) heat gain through windows and skylights

## Heat losses

The main heat losses are the following:

*Heat losses though the interfaces of the dwelling*
Heat is lost through walls, windows, floor and ceiling, by conduction and radiation. These are calculated individually by considering each area separately. Details of how such heat losses can be calculated are given on page 36.

*Heat losses due to ventilation*

It is first of all necessary to calculate the volume of air in $m^3$

which is being changed each second. This is equal to:

$$\frac{\text{airchange rate} \times \text{volume of house}}{3600}$$

The specific heat of air at 20°C equals 1·21 kJ/m$^3$ K
Therefore the heat loss in joule/second or watt equals volume of air in m$^3$ × 1·21 × difference between internal and external ambient temperatures.

WORKED EXAMPLE

A house has an internal volume of 250 m$^3$ and the air change practised is 1·2 per hour. The internal ambient temperature is 20°C and the external ambient temperature is 10°C. Work out the ventilation heat losses.

$$\text{Volume of air lost per second} = \frac{250 \times 1{\cdot}2}{3600} = 0{\cdot}083\ \text{m}^3$$

$$\text{Heat loss} = 0{\cdot}083 \times 1{\cdot}21 \times 10 = 1{\cdot}01\ \text{kJ/s}$$
$$= 1{\cdot}01\ \text{kW}$$

*Heat losses due to run-off of water*

Water has a specific heat of 4·187 kJ/kg K
Each litre of water run-off therefore represents:
4·187 × (temperature of hot water − temperature of cold water) kJ
If the hourly run-off equals, say, 20 litres of water at 70°C, while the town's water comes in at 15°C, the run-off losses equal:

$$\frac{20 \times 55 \times 4{\cdot}187}{3600}\ \text{kJ/s} = 1{\cdot}28\ \text{kW}$$

## Dimensioning of a heating system

To find out what size heating system is required, one first of all assumes a figure for the lowest external temperature likely to be experienced for any length of time. This is not as easy as it sounds. According to the *Guinness Book of Records*, London on the 5th January 1867 had a temperature of −17°C. One cannot, however, design a heating installation according to such an unlikely occurrence, which could happen once a century. Normally temperatures below −5°C are rare anywhere in the British Isles and this would be a reasonable bottom figure. West of the Pennines one could standardise on a slightly higher figure than this, i.e. about $-3°C$, while in high-lying or exposed areas one could assume a slightly lower figure, say $-8°C$. For the internal ambient temperature figure one takes the optimum comfort temperature desired. As it is quite likely that the dwelling may at some time be occupied by people who have higher comfort demands than the average, an internal ambient temperature assumption of 25°C would not be unreasonable.

Using therefore the two ambient temperatures mentioned as a base, one calculates:

(a) conduction and radiation heat losses
(b) ventilation heat losses
(c) water run-off heat losses

and expresses the sum of these in kilowatt.
From this figure one then deducts the sum of:

(a) heat gain due to electrical equipment
(b) heat gain due to occupancy
(c) heat gain through windows and skylights

again expressing this figure in kilowatt.

The capacity of the heat appliances needed equals the difference between heat losses and heat gains.

If the heat gains during part of the year should be greater than the heat losses, one would need an air conditioning plant to remove this excess heat. This is normally the case in the United Kingdom in office buildings, department stores, etc., but not to any extent in private dwellings, except during comparatively short spells of hot weather. Air conditioning is therefore not required in domestic premises in the United Kingdom.

**How much ventilation?[5]**

When human beings breathe in air, they absorb the oxygen contained in it and expire carbon dioxide. One can breathe in air with up to 3–4% carbon dioxide without any ill effect whatever, or even without detecting the fact. The real criterion as to how much ventilation is needed is dependent upon other factors. These are:

(a) the dissipation of water vapour produced in kitchen and bathroom
(b) the dissipation of smells

In living rooms, where little additional water vapour is being produced, one needs about 17 m³ of fresh air per person an hour to keep the air reasonably fresh. In contrast, even in very crowded and stuffy rooms the carbon dioxide content is hardly ever more than 0·1–0·2%, well within the normal tolerance limit of 3–4%.

In other rooms where water vapour is being produced as well, one may have to aim at a somewhat higher ventilation rate. But the Institution of Heating and Ventilating Engineers accepts a figure of 17 m³ per person an hour as a reasonable design criterion for dwellings. In old-fashioned houses where windows and doors fit badly, air change rates of up to two per hour exist without any artificial means of ventilation being present. In addition, an old-fashioned open coal fire draws out about 200 m³ of room air each hour, which is replaced from the outside. Over-ventilation rather than under-ventilation is usually the problem with old houses. In more modern houses, where better standards of workmanship are followed, the natural ventilation rate is about one air change per hour. In addition, if one uses a heating system which does not act as a ventilation system as well, natural air change is the only means relied upon to provide the ventilation for removal of body odours, cigarette smoke and so on. It is easy to work out whether a living room is adequately ventilated or not. One simply multiplies its length by its width and height to obtain its volume in m³. If the figure obtained is then divided by 17, the result gives the maximum number of people who should use the living room without requiring additional ventilation.

WORKED EXAMPLE

A room is 6 metres long, 4·5 metres wide and 3 metres high. For how many people is natural ventilation adequate?

The volume of the room is $6 \times 4{\cdot}5 \times 3 = 81$ m³

Natural ventilation is adequate if the room is being used by

$$\frac{81}{17} = \text{just over four adults}$$

If more people than this are present, the room is likely to get stuffy.

Rooms are also likely to get stuffy if the air inside them does not move fast enough. The average air speed inside a dwelling should be about 200 mm/s. Air speeds below 100 mm/s feel stuffy, while speeds in excess of 500 mm/s feel draughty.

The *relative humidity* of a dwelling is also of importance. If it exceeds 65% the air inside the room feels sticky and uncomfortable. Better ventilation serves to lower such high humidity rates. Relative humidities below 30% are also unpleasant. Permanent exposure to such low humidities can cause sore throats and coughs. Too low a relative humidity, which is often found in very warm centrally heated dwellings, can be counteracted by the installation of a humidifier. The simplest humidifiers are vessels filled with water, which are hung on to radiators. There are far more sophisticated pieces of equipment around which sense the relative humidity of the air and automatically adjust the water vapour content of it to a pre-selected figure.

## Methods of heating

### *Individual heating*

The following methods of individual heating are currently used in the United Kingdom:

*1 Open coal fires*
These are built into either the external wall or into an internal wall. These traditional open fireplaces are still widely found in old dwellings, but are not being installed into new houses. The disadvantages are dirt, the labour of lighting them, heavy air-pollution problems and extremely low efficiency. Thermal efficiencies vary between a maximum value of about 25% down to a lower limit of only 15%. Such heating appliances are only permitted in smoke-controlled areas if smokeless fuel is used in them. As smokeless fuel is rather more expensive than raw coal and also has a lower calorific value, smokeless open coal fires are even more expensive to run than smoky ones.

*2 Modified solid fuel combustion appliances*
These are mainly designed to burn smokeless fuel such as anthracite, processed coal and coke. They have efficiencies varying from about 30% to about 60%. Often they are connected to one or more water-filled radiators. The capital cost of such appliances is much lower than that of central heating. The running cost is, however, much higher.

*3 Paraffin heaters*

Paraffin heaters operate at a very high efficiency, and because the fuel is at present a good deal cheaper than competitive fuels, paraffin heaters are popular with low-income families. When flues are fitted, paraffin heaters are totally satisfactory. Unfortunately many paraffin heaters are installed without flues. Each litre of paraffin, when burned, produces approximately 1·5 $m^3$ of carbon dioxide and 1·3 kg of water vapour. Also, if the air supply is restricted, it can happen that extremely poisonous carbon monoxide may be produced. At its best flue-less paraffin heaters can induce heavy condensation on walls, windows and ceilings. At their worst they can cause serious damage to health.

Paraffin burns in heaters at a very high efficiency, estimated to be about 75%. The capital cost of the heaters is low, as is the price of the fuel. Under existing economic conditions this form of heating is by far the cheapest.

*4 Electric fires*

Electric fires are cheap to buy, cheap to install and run at 100% efficiency. Their advantage is also complete flexibility, absence of smell, fumes, etc. The disadvantage of electric fires, which use on-peak electric current, is the cost of power. Even the most modern power stations are unable to convert their primary fuel into electricity at an efficiency much in excess of 36%, while the overall average conversion rate of fuel into power currently achieved by the CEGB is around 27%. In addition, power generation requires an immense capital investment in generating plant, switchgear, transmission lines, etc. and high labour and administration costs. For this reason the cost of electric power is of necessity several times as high per unit heat contained in it than a primary fuel such as oil, coal or gas. Running costs for on-peak electric fires are therefore very high, but the high costs are somewhat relieved by the fact that one can switch an electric fire off the moment one no longer requires it.

*5 Gas fires*

Modern gas fires operate at a very high thermal efficiency. They are reasonably cheap to buy and to install, and in consequence are very popular. Like electric fires they can be switched on and off instantaneously and give off their heat both by radiation and by convection. Gas fires must be fitted to a flue and in consequence are not quite as flexible to install as electric heaters. Fuel costs are, however, much lower.

*Central heating systems*[6]

The following are the main classifications of central heating systems:

*1 Off-peak electric systems*

Power stations could be run most efficiently if electricity could be supplied evenly throughout the 24-hour period. Unfortunately, demand varies enormously over the day. It is maximum in the morning, when factories start up, and again during the evening, falling off to a very low figure during the night. To encourage more even demand, the electricity authorities offer power at approximately half-price during the off-peak periods. A special meter (white) is used for off-peak current. The electricity is used at night to heat up either floors of buildings or special storage cores, the heat being given off during the day. It is difficult to store enough heat in floors to provide adequate heating for a dwelling. Under-floor heating can be regarded purely as background heating as loads in excess of 200 $W/m^2$ would make the floor too hot for comfort. In addition, it is necessary to practise highly efficient under-floor thermal insulation, otherwise large heat losses would make the system uneconomic. Electric storage heaters, particularly when fitted with adjustable fans to control heat emission, are economic to run and constitute a highly competitive system of heating.

*2 Water-filled systems*

Gravity operated systems are now becoming obsolete and the modern water-filled narrow-bore systems use an electric motor to pump the circulating water around. The systems can use either solid fuel, oil, gas or off-peak electric current for heating up the water. In Europe perhaps the most popular way of operating such systems is by means of district heat, i.e. using water heated up in central stations by means of waste heat from electric power stations, refuse incinerators, industrial processes or large scale furnaces using cheap grades of oil or coal. Such district heating systems are now being introduced into Great Britain too, although it is likely to take many years before they will be at all common.

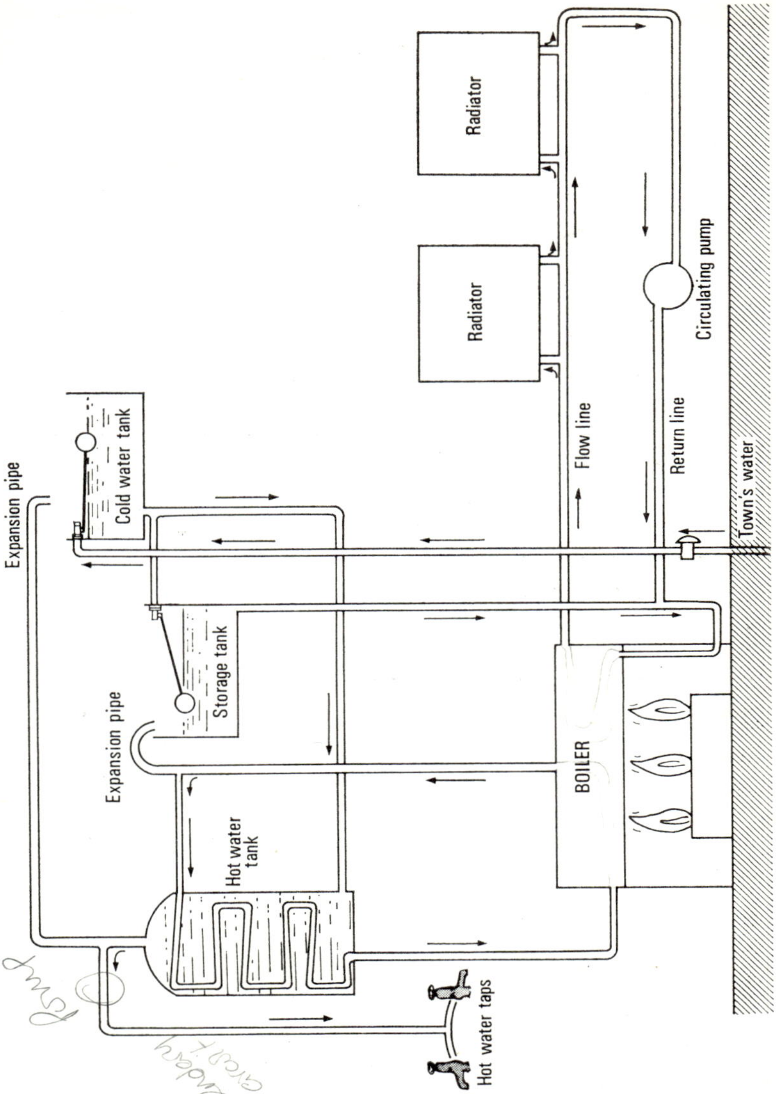


FIG. 2 Layout of water-filled central heating system

The advantages of hot water systems are:

(a) they can easily be installed into existing dwellings
(b) they are quiet in operation
(c) they are foolproof in operation

Their disadvantages are mainly that they are slow to respond to changes of climatic conditions as it takes a certain time for the water and the pipes in which it is carried to heat up. The radiators take up space and are aesthetically unpleasing.

Various types of radiators are used, with different output ratings. The output rating of different radiators varies enormously. The following are approximate figures for different designs of radiators.

TABLE 5. Output of different types of radiator sunder normal conditions[25]

| *Type of radiator* | *Output (kW/m²)* |
|---|---|
| Fan convector | 8·5 |
| 5-column radiator | 4·5 |
| 3-column radiator | 3·0 |
| Corrugated radiator | 2·5 |
| Louvred radiator | 2·5 |
| Double-panel radiator | 2·0 |
| Single-panel radiator | 1·0 |

In general, radiators which have a low output rating are much more satisfactory in use than high output units such as fan convectors which will cause heat to be distributed unevenly in the room. Radiators should be fixed low down, preferably underneath windows. Room air is then cooled by the window and drifts downward to be heated by the radiator. This changes its flow to a sideways motion, giving up some of the heat to the floor, which is able to radiate it to the occupants.

Radiators which are placed high up tend to produce uneven distribution of heat. The parts underneath the ceiling are then the warmest sections of the dwelling, while areas just above the floor are cold. It is possible for temperature differences as high as 5°C to occur under such circumstances. Low floor temperatures produce discomfort even if the general room temperature is high.

Skirting heaters have been found to be particularly successful. The output rating of the four most popular types of such heaters is as follows:

| | |
|---|---|
| cast iron skirting (200 mm deep) | 0·25 W/m |
| low panel skirting | 0·5 W/m |
| finned pipe skirting | 0·4 W/m |

A system which is now becoming increasingly popular is to circulate hot water in ceiling panels. Alternatively the ceiling panels are heated electrically. A surface temperature of about 40°C is usually employed and this serves to distribute the heat in the dwelling by radiation. Unlike floor heating, where the main heat transfer is by convection which may cause uneven distribution of heat, ceiling systems which radiate the heat out to the walls, floors, furniture and other parts of the dwelling give out even heat. It is, however, of the utmost importance to employ really good insulation in the ceiling section, which is placed above the heaters.

*3 Hot air systems*

These are systems where cold air is drawn in from the room, sometimes mixed with a certain quantity of external air, and blown out through registers after having been heated up in a central unit which may be either solid fuel fired, oil-fired or gas-fired. Registers can be placed either in walls or into the floor. The main advantages of hot air systems are that they have a very quick response and, once lit, heat up a dwelling much quicker than water-filled radiator systems. In addition, the systems can be adjusted to provide additional ventilation. Although efficient dirt filtration units are normally fitted to hot air systems, they do tend to cause dirt deposition above the registers. The air ducts have a large cross-sectional area and are often insulated. In consequence a hot air system is very difficult to fit into an existing dwelling. Finally, there have been complaints about the noise of the driving fan, although the author thinks that such complaints are somewhat unjustified.

*Comparative capital costs*

It is difficult, during a period of galloping inflation, to give cost figures which do not look quite ridiculous a few years afterwards. The figures given apply to central heating systems with a capacity of 7 kW, and are obtained from the Consumers' Association publication dealing with central heating. They applied during the year 1970. In future years they may serve as comparative rather than as actual figures.

| | |
|---|---|
| Off-peak electric heating using storage heaters | £192 |
| Direct electric heating systems with oil-filled radiators | £220 |
| Gas-fired warm air system | £247 |
| Small bore gas-fired system | £270 |
| Hand fed coke-fired boiler and small-bore system | £286 |
| Hopper fed coke-fired boiler and small-bore system | £330 |
| Small bore oil-fired system | £380 |
| Hot air oil-fired system | £396 |

The oil-fired systems are more expensive because of the need to install a fair-sized oil tank.

*Comparative fuel costs*

Fuel prices are even more difficult to compare than capital costs. They vary in different parts of the country, complex scales of varying tariffs are used, and of course the units employed and efficiency of combustion are not the same for different fuels.

*1 Solid fuels*

These are sold by the imperial hundredweight (50·802 kg). The calorific value of the solid fuel supplied is a very critical factor.

Dry anthracite has a calorific value of 1·7 GJ per hundredweight and can be burnt at 75% efficiency. This means that in a modern furnace one can obtain 1·28 GJ from a hundredweight of this fuel.

Dry coke has a calorific value of 1·4 GJ per hundredweight and can be burnt at 65% efficiency. This means that in a modern furnace one can get 0·91 GJ from each hundredweight of coke. If the fuel supplied is damp, figures many per cent lower than these will be obtained. It is not uncommon for coke which has been left out on a heap in the rain to contain 30% water. Apart from the fact that water does not burn, considerable heat expenditure is necessary to convert the water into steam. One needs roughly 2·5 MJ to convert a kilogramme of water into steam.

WORKED EXAMPLE

Damp coke contains 30% water. Determine the useful heat present in a hundredweight (50·802 kg) of this fuel.

The dry coke contains 0·91 GJ net heat. If it is wet it contains:

$$0{\cdot}7 \times 0{\cdot}91 - \frac{(50{\cdot}802 \times 0{\cdot}3 \times 2{\cdot}5)}{1000} \text{ GJ}$$
$$= 0{\cdot}60 \text{ GJ}$$

If the cost of a hundredweight of the solid fuel is known, one can then divide this by the number of effective GJ obtainable from such a weight of fuel, to determine the actual cost per GJ of heat. (1 GJ = 1000 MJ = 1 000 000 kJ = 1 000 000 000 J).

*2 Liquid fuels*

Liquid fuels are commonly sold by the imperial gallon (4·546 litres). An imperial gallon of kerosene (paraffin) has a calorific value of 0·168 GJ and can be burnt with 75% efficiency. The heat obtainable from a gallon of paraffin is therefore 0·125 GJ or 125 MJ. The grade of oil which is used in domestic central heating installations has a calorific value of 0·175 GJ per gallon and can also be burned with an efficiency of 75%. The heat obtained from a gallon of oil is therefore 0·131 GJ. If the price per gallon is known, one can easily evaluate the cost per GJ.

*3 Gaseous fuels*

Two types of gas are currently being supplied to British householders. These are: normal town's gas with a calorific value of about 18 $MJ/m^3$ and natural gas (North Sea gas) with a calorific value of 35 $MJ/m^3$. Both give excellent results in combustion appliances but their burning properties are so different from each other that equipment designed for normal town's gas cannot be used for burning natural gas and vice versa. Conversion from normal town's gas to natural gas is possible if carried out by highly skilled mechanics but results so far have not been too encouraging.

Charges for both types of gas are made per therm, where a therm equals 0·105 GJ. Gas burns at an efficiency of 75%, so that the therm really equals 0·079 GJ or 79 MJ actual heat obtained.

Normally town's gas is sold either with a flat charge and a rate of so much per therm, or on a sliding scale which is worked out in such a way that the more one buys the lower the price per therm is.

To work out the actual cost of running a central heating system using gas is thus rather complex. If no other gas-burning appliance is used, one simply takes the entire monthly bill and divides the sum of money by the number of therms consumed. This gives the cost per therm and hence the cost per useful GJ. But if other gas appliances are used as well, one has to calculate how much the bill would have been without the gas consumed by the central heating appliance, and subtract this from the total bill. This cost is then divided by the number of therms used for central heating.

WORKED EXAMPLE

212 therms of gas are used, of which 45 are consumed by the gas cooker and the rest by the central heating system.
The following charging system is used:

First 15 therms at 15p/therm
Next 100 therms at 12p/therm
Next 500 therms at 10p/therm

Calculate the cost of gas for central heating purposes only in p/GJ.

The central heating appliance uses 212 − 45 = 167 therms, of which 70 therms are charged at 12p and 97 at 10p.

The total cost of the 167 therms used is thus:

$$\begin{aligned} 70 \times 12\text{p} &= £\ 8{\cdot}40 \\ 97 \times 10\text{p} &= £\ 9{\cdot}70 \\ &\overline{£18{\cdot}10} \end{aligned}$$

The cost per therm is thus $\frac{£18{\cdot}10}{167} = 10{\cdot}83\text{p}$

The cost per useful GJ is therefore:

$$\frac{10{\cdot}83\text{p}}{0{\cdot}079} = £1{\cdot}37$$

*4 Electric current*

Electric current, whether on-peak or off-peak, is always measured in units of kilowatt hours.

1 kWh = 3·6 MJ, or 278 units of electricity equal 1 GJ in heating value. As electricity can be converted at 100% into heat one does not have to obtain the useful heat figure as with other kinds of fuels. 278 units of electricity always equal 1 useful GJ.

Very much the same kind of charging procedure exists with

electric power as with gas. There is either a standing charge, or a certain number of units are initially charged for at a higher rate. To evaluate the fuel cost for an electric heating system one has to again subtract the normal power requirements for lighting, cooking etc. from the total bill, and then equate the additional cost with the number of units used exclusively for heating.

To find the total heating cost per useful gigajoule, which is extremely important in calculations involving economic insulation, we must know the following data:

| | |
|---|---|
| (a) cost of the heating appliance and auxiliaries in £ | $A$ |
| (b) interest rate to be paid on the capital | $r$ |
| (c) expected life of the installation | $p$ |
| (d) annual maintenance and repair costs in £ | $R$ |
| (e) cost of the fuel per useful GJ in £ | $F$ |
| (f) annual production of heat in GJ | $Q$ |

The cost per useful GJ is then the following:

$$\pounds\ \frac{A\left(\frac{r}{100}+\frac{1}{p}\right)+R}{Q}+F$$

WORKED EXAMPLE

The cost of the appliance is £350 and it is expected to last 25 years. The interest payable is 7% and the annual repair and maintenance costs are estimated at £4. Oil is used as a fuel and this is found to cost 7·3p per gallon. Oil has a useful heat content of 0·131 GJ per gallon so that the fuel cost $F$ comes to

$$\pounds\frac{0{\cdot}073}{0{\cdot}131}=\pounds 0{\cdot}557$$

The annual heat production is 41·4 GJ
The total cost per GJ therefore amounts to:

$$\pounds\frac{350(0{\cdot}07+0{\cdot}04)+4}{41{\cdot}4}+0{\cdot}557$$
$$=\pounds 1{\cdot}125+0{\cdot}557=\pounds 1{\cdot}682$$

The cost per useful MJ is one thousandth of this figure, namely:
0·17p

# 3 Thermal Characteristics of Building Elements

## Calculation of heat losses through walls and windows

The heat loss by conduction and radiation through a solid interface which may consist of one or more layers is always expressed by the following formula:

$$Q = U \times (t_i - t_o) \times A$$

where $Q$ is the rate of heat lost in watt

$U$ is the so-called $U$-value of the interface in question which is given in SI units as follows: watt/metre$^2$ K or watt/metre$^2$°C

$t_i$ is the internal ambient temperature in °C

$t_o$ is the external ambient temperature in °C

$A$ is the area of the wall or other interface in m$^2$

WORKED EXAMPLE

The internal temperature is 20°C, the external temperature is 10°C and the area of the wall is 60m$^2$. If the $U$-value of the wall is 1·7 W/m$^2$ K (the $U$-value of the traditional 11-inch cavity brick wall in SI units), the rate of heat loss equals:

$$1{\cdot}7 \times 60 \times 10 = 1020 \text{ W} = 1{\cdot}02 \text{ kW}$$

This means that each second 1·02 kJ of heat are lost

*How a* U*-value is obtained*[7,8]

In any wall, the $U$-value depends on three main features:

(1) a boundary layer of air at the inside of the dwelling
(2) the nature of the wall itself, i.e. the type of materials used in

its construction, its thickness, any enclosed air spaces, foil reflectors, and so on

(3) a boundary layer of air at the outside

While the thickness of the boundary layer of air in contact with the inside wall surface is virtually constant, that of the outside varies enormously with the wind which passes along the wall. This is a matter which is often ignored when *U*-values of walls and windows are being calculated. In the case of single-glazed windows, the glass in the window has a completely negligible heat resistance.

The only resistance to heat transfer is in the two air layers, one on the inside and one on the outside of the glass. One can say that in all cases where there is a thin lamina of material only, like a sheet of metal, a stretched film of polyethylene foil or even a thin sheet of asbestos cement, the two air layers constitute the entire thermal resistance. Only if the wall itself has a substantial thermal resistance does one include its thickness in the calculation of the *U*-value.

To calculate the *U*-value of a wall, one uses a reciprocal formula of the type used by electricians to calculate electrical resistances in parallel.

This formula is:

$$\frac{1}{U} = \frac{1}{h_i} + \frac{1}{h_w} + \frac{1}{h_o}.$$

where $U$ is the overall heat-transfer coefficient of the $U$-value in watt/metre$^2$ K

$h_i$ is the internal boundary layer heat-transfer coefficient

$h_w$ is the heat-transfer coefficient of the wall

$h_o$ is the external boundary layer heat-transfer coefficient

$h_i$, $h_w$ and $h_o$ are all given in watt/metre$^2$ K.

*Calculation of the two boundary layer heat-transfer coefficients*

The value of the internal boundary layer heat-transfer coefficient is always virtually the same, namely

$$h_i = 8{\cdot}5 \text{ W/m}^2 \text{ K}$$

On the other hand, the external boundary layer heat-transfer coefficient varies widely, according to the following factors:

(a) wind speed

(b) surface roughness of the material

Table 6 gives the $h_o$ values for a number of different surfaces under different wind speed conditions.

TABLE 6. $h_o$ values in $W/m^2$ K[7]

| *Wind speed parallel to surface in metres/second* | *Nature of surface* | | | | |
|---|---|---|---|---|---|
| | *Glass or metal* | *Plaster or timber* | *Concrete* | *Brick* | *Stucco* |
| Wind still | 8·5 | 8·5 | 8·5 | 8·5 | 8·5 |
| 2 | 12·8 | 13·5 | 14·6 | 15·2 | 16·8 |
| 5 | 26·5 | 32·8 | 42·4 | 47·6 | 61·0 |
| 10 | 46·5 | 55·7 | 74·5 | 83·5 | 107·0 |
| 15 | 63·9 | 76·6 | 102·2 | 115·0 | 147·0 |
| 20 | 81·3 | 97·6 | 130·5 | 146·6 | 187·5 |

FIG. 3 Aerated concrete roofing slabs combine excellent strength characteristics with good thermal insulation properties

*Calculation of $h_w$ for walls made of a single material*

For walls which are made from a single material only, such as solid brickwork, solid gas (aerated) concrete, concrete, timber and so on, it is an easy matter to determine the $h_w$ value. It is simply:

$$\frac{\text{the } k\text{-value according to Table 7}}{\text{the thickness of the wall in metres}}$$

Using the SI nomenclature the thermal conductivity of materials is now given in terms of:

W/m K

I think that everyone must agree that this is a great improvement over the old system of units when *k*-values were given usually in:

inch Btu/hour foot$^2$ °F

Table 7 gives the *k*-values of a number of solid building materials in the new SI units.

TABLE 7. *k*-values of building materials in W/m K[9]

| | | | |
|---|---|---|---|
| Aluminium | 198 | Copper | 380 |
| Lead | 34·6 | Steel | 48·3 |
| Zinc | 112 | Aerated concrete (400 kg/m$^3$) | 0·086 |
| Asbestos cement | 0·75 | Aerated concrete (500 kg/m$^3$) | 0·103 |
| Breeze block | 0·17–0·35 | Aerated concrete (650 kg/m$^3$) | 0·138 |
| Brickwork | 0·7 | Aerated concrete (800 kg/m$^3$) | 0·210 |
| Concrete | 0·7 –1·9 | Glass | 0·5 –0·8 |
| Plaster of Paris | 1·09 | Timber about | 0·2 |
| Rubber | 0·2 | Vermiculite concrete between | 0·08–0·2 |

WORKED EXAMPLES

(1) Determine the effective *U*-value of a wall of gas concrete of density 500 kg/m$^3$, which is 200 mm thick and is exposed to a wind parallel to its surface of 10 metres/second.

Looking at Table 7 we obtain a *k*-value of 0·103 for the gas concrete in question. Its thickness is 0·2 metres, so that $h_w$ is equal to $\frac{0{\cdot}103}{0{\cdot}2} = 0{\cdot}515$ W/m$^2$ K

$h_i$ we know—it is 8·5 W/m² K

$h_o$ we can find from Table 6. Gas concrete has roughly the same surface roughness as brickwork, so that we can assume $h_o$ to be 83·5 W/m² K

Using our reciprocal equation we may now write:

$$\frac{1}{U} = \frac{1}{8{\cdot}5} + \frac{1}{0{\cdot}515} + \frac{1}{83{\cdot}5}$$

$1/U$ is therefore equal to 2·07
and $U$ is equal to 0·483 W/m² K

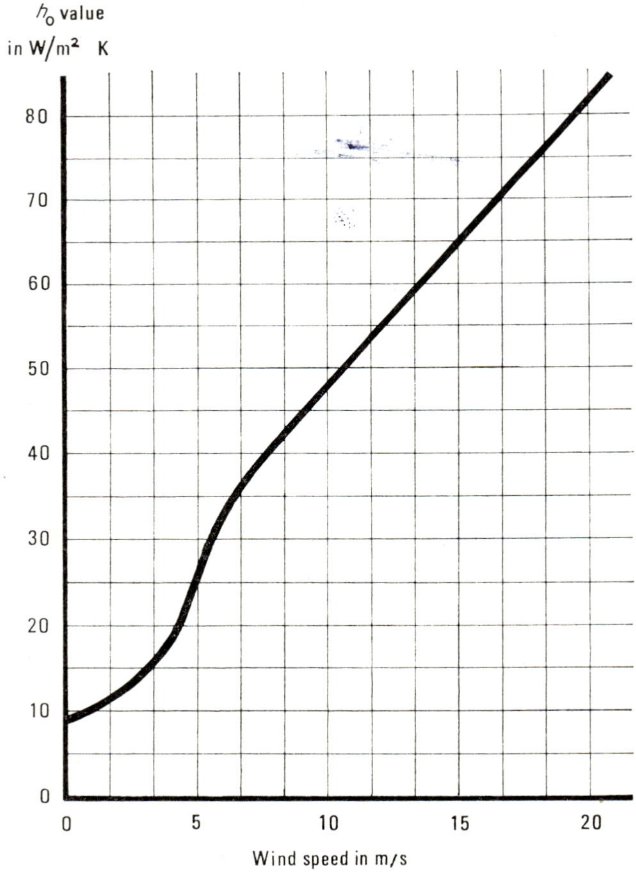


FIG. 4 Graph showing relationship between wind speed and $h_o$ for smooth surfaces

(2) Determine the $U$-values of single glazed windows under completely wind-still conditions, and also when a 20 metres/second wind is blowing. In this case $h_w$ can be completely ignored as it is such a large value that its reciprocal will be zero.

Therefore:

$1/U = 1/h_i + 1/h_o$

Under wind-still conditions both $h_i$ and $h_o$ are 8·5 W/m² K; therefore $1/U = 0{\cdot}235$ and $U = 4{\cdot}25$ W/m² K

If a wind of 20 metres/second is blowing, then we can take $h_i$ to be 8·5 W/m² K as before, but $h_o$ is now 81·3 W/m² K

We can write our formula as:

$1/U = 1/8{\cdot}5 + 1/81{\cdot}3$ or

$1/U = 0{\cdot}129$

$U$ is therefore equal to 7·75 W/m² K

It can be seen that the $U$-value of a window under conditions of strong wind is virtually double that of the same window under wind-still conditions.

In the past, it was usual to quote the $U$-values of a wall or window as an arbitrary figure. As can be seen, the wind exposure makes complete nonsense of such figures, especially if the wall is badly insulated.

Wind speeds for given areas and at different building heights are generally measured by an anemometer, and will be readily given to anyone who cares to enquire at the local meteorological station. Using the maximum values likely to be encountered in a given area and at a given building height in the equations given, the architect will be able to calculate the maximum heat loss likely to be encountered by his building. For bigger contracts it is probably even worth while to carry out wind-tunnel tests, or at any rate attempt to calculate how shading by hills, trees and so on is likely to affect the wind speed past façade surfaces.

*More complicated walls*

Let us now consider the calculations involved when more complicated walling constructions are involved.

Whenever a wall consists of a number of parallel layers, we must again use our reciprocal formula for calculating the heat transfer through the wall.

$h_w$ is the heat transfer through the wall alone, without taking into consideration the internal and external boundary layers of air.

A common method of construction today is to use twin solid structural leaves of brickwork or concrete, and sandwiching a layer of either foamed plastic or mineral wool.

Table 8 gives the $k$-values of some of these insulation materials, again, of course, in the SI units of W/m K.

TABLE 8. $k$-values of insulation materials in W/m K[9]

| | |
|---|---|
| Loose diatomaceous earth | 0·06 |
| Diatomite | 0·12 |
| Expanded polystyrene | 0·035 |
| Fibre insulation board | 0·049 |
| Glass fibre quilt | 0·043 |
| Hair | 0·036 |
| Polyurethane foam | 0·028–0·043 |
| Urea formaldehyde foam | 0·036 |
| Cork | 0·043 |

WORKED EXAMPLE

Determine $h_w$ of a typical sandwich wall of the type frequently used in industrialised building, which consists of:
76 mm of concrete, 32 mm of expanded polystyrene and 76 mm of concrete.
The $k$-value of concrete is equal to 1·0 W/m K for the inner leaf and 1·6 W/m K for the outer leaf. The value for the expanded polystyrene can be obtained from Table 8 and is 0·035.
Using the reciprocal formula:

$$1/h_w = a_1/k_1 + a_2/k_2 + a_3/k_3 \ldots \text{ etc.,}$$

where $a_1$, $a_2$, $a_3$ etc. are the thicknesses of the various layers in metres
and $k_1$, $k_2$, $k_3$, etc., are the $k$-values of the materials which make up the various layers, we can obtain the following sum:

$$1/h_w = 0{\cdot}076/1{\cdot}6 + 0{\cdot}032/0{\cdot}035 + 0{\cdot}076/1{\cdot}0$$
$$= 0{\cdot}0475 + 0{\cdot}915 + 0{\cdot}076 = 1{\cdot}039$$

or $h_w = 0{\cdot}965$ watt/metre² K

To determine the $U$-value of this wall, we must know the wind speed past the surface.

$h_i$, the inner boundary layer heat-transfer coefficient, is always 8·5 W/m² K, while the external boundary layer heat-transfer coefficient varies between

8·5 W/m² K for wind-still conditions to

42·4 W/m² K for winds of 5 metres/second to 130·5 W/m² K for winds of 20 metres

To be accurate the $U$-value is calculat
as follows:

$1/U = 1/8{\cdot}5 + 1/0{\cdot}965 + 1/8{\cdot}5 = 1{\cdot}274$ so that
$U \quad = 0{\cdot}785$ W/m² K for wind-still conditions
$1/U = 1/8{\cdot}5 + 1/0{\cdot}965 + 1/42{\cdot}4 = 1{\cdot}181$ so that
$U \quad = 0{\cdot}835$ for winds of 5 m/s
$1/U = 1/8{\cdot}5 + 1/0{\cdot}965 + 1/130{\cdot}5 = 1{\cdot}165$ so that
$U \quad = 0{\cdot}860$ for winds of 20 m/s

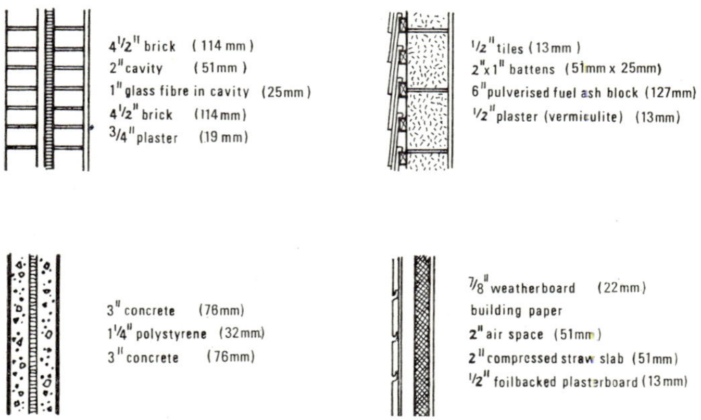


FIG. 5 (a) Walls with $U$-value 0·57 W/m² K

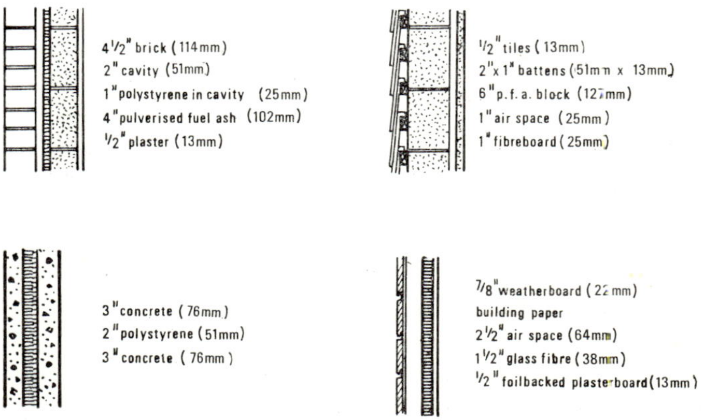


FIG 5 (b) Walls with $U$-value 0·85 W/m² K

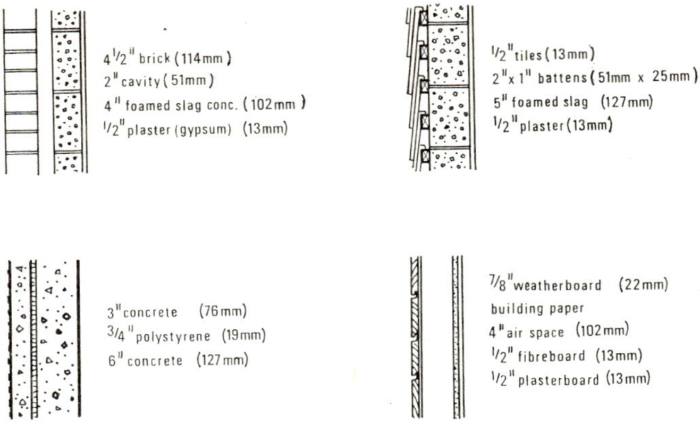


FIG. 5 (c) Walls with $U$-value 1·13 W/m² K

The Ministry of Housing and Local Government has published a design bulletin entitled *Space and Water Heating 3, Part 8* (1967).[10] $U$-values are quoted in this for a number of different walling constructions, but no mention is made of the effect of the wind speed upon these values. When the values are recalculated from first principles, it becomes obvious that the assumption was being made in all cases, that the wind speed past the external façade was of the order of 2–3 metres/second as the $h_o$ values came out as 15–25 W/m² K. The design bulletin gives all values in British units, but as it is essential that builders and architects must start to think metric now, as in a few years' time metric units will be the only ones generally used, their data has been converted into SI units in Table 9. These figures apply to wind speeds of between 2 and 3 metres/second. For different wind speeds the values must be loaded according to Table 10.

TABLE 9. $U$-values quoted by the Ministry of Housing and Local Government for varying wall constructions. All $U$-values are given in W/m² K[10]

| | |
|---|---|
| 114 mm brick, 51 mm cavity, 102 mm foamed slag, 13 mm gypsum plaster | $U$=1·13 |
| 114 mm brick, 51 mm cavity, 25 mm glassfibre, 114 mm brick, 19 mm plaster | $U$=0·85 |
| 114 mm brick, 51 mm expanded polystyrene, 102 mm pulverised fuel ash block, 13 mm plaster | $U$=0·57 |

| | |
|---|---|
| 76 mm concrete, 19 mm expanded polystyrene, 152 mm concrete | $U=1{\cdot}13$ |
| 76 mm concrete, 32 mm expanded polystyrene, 76 mm concrete | $U=0{\cdot}85$ |
| 13 mm tiles, 51 mm × 25 mm battens, 127 mm foamed slag blocks, 13 mm plaster | $U=1{\cdot}13$ |
| 13 mm tiles, 51 mm × 25 mm battens, 152 mm pulverised fuel ash blocks, 13 mm vermiculite plaster | $U=0{\cdot}85$ |
| 22 mm weatherboard, building paper, 102 mm air space, 13 mm fibreboard, 13 mm plasterboard | $U=1{\cdot}13$ |
| 22 mm weatherboard, building paper, 51 mm air space, 51 mm compressed straw slab, 13 mm foil-backed plasterboard | $U=0{\cdot}85$ |
| 13 mm tiles, 51 mm × 25 mm battens, 152 mm pulverised fuel ash blocks, 25 mm air space, 25 mm fibreboard | $U=0{\cdot}57$ |
| 22 mm weatherboard, building paper, 64 mm air space, 38 mm glass fibre, 13 mm foil-backed plasterboard | $U=0{\cdot}57$ |
| 230 mm solid brickwork, 13 mm plaster | $U=2{\cdot}26$ |
| 115 mm brick, 51 mm air space, 115 mm brick, 13 mm plaster | $U=1{\cdot}7$ |

TABLE 10. The percentage loading of the values given above which must be carried out under different wind speed conditions

| | |
|---|---|
| Wind still | −10% |
| 2 metres/second | 0% |
| 5 metres/second | + 5% |
| 10 metres/second | + 7% |
| 15 metres/second and more | +10% |

These figures are very approximate. If accurate figures should be required it is necessary to calculate them from first principles.

*The effect of air cavities in walls*

Many walls include air cavities in their make-up. These air cavities serve two purposes:

(1) they act as a moisture stop to prevent water passing from the outside to the inside by capillary action.
(2) they act as thermal insulation layers

As far as the first action is concerned, this will be dealt with later. We are now concerned with the effect of an air cavity as an insulation layer. On the face of it, air is the best insulator one can find. Its $k$-value is equal to 0·028 W/m K. Only gases such as argon, carbon dioxide, etc. have lower $k$-values. In fact,

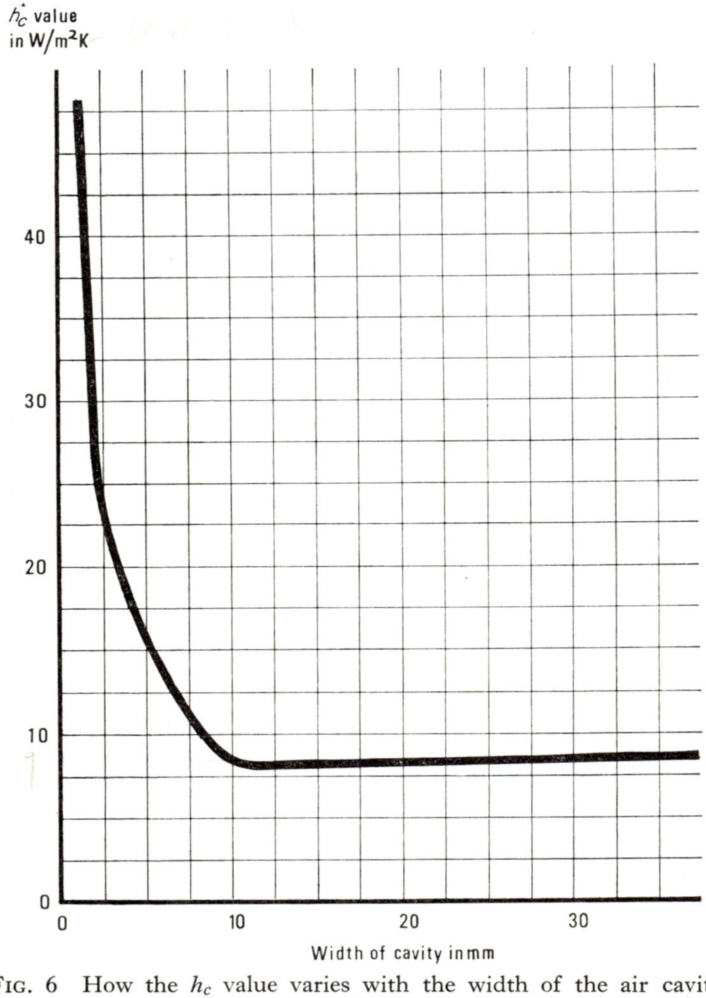


FIG. 6 How the $h_c$ value varies with the width of the air cavity employed (Brickwork with ventilated cavity running from top to bottom of house)

it is no coincidence that all insulation materials which consist mainly of voids such as foamed plastics, mineral wool and glass wool should have virtually the same $k$-values, which are roughly 20% higher than that of air itself. The high void insulators, as they are called, are in fact nothing but still air, with bits of material round them.

In the case of an air cavity, the only resistance to heat flow is in relatively thin layers of air, which are attached to the two faces of material bordering the cavity. In the centre of the cavity, the air moves freely and carries the heat with it by convection. We can say that, with an air gap between 0 and 10 mm in width, the $1/h_c$ value ($h_c$=heat transfer value due to cavity in W/m² K) goes up from 0 to 0·146 and then stays around this figure, however much the air gap is then widened. This means that $h_c$ for an air gap in brickwork, concrete, timber and so on is always 6·8 W/m² K, provided the gap is more than 10 mm wide.

*Effect of covering one side of an air gap with a shiny material such as aluminium foil*

The use of aluminium foil to face one side of a cavity has the effect of cutting out all radiation heat transfer across the cavity. This reduces the $h_c$ value to 2·3 W/m² K as long as the aluminium remains shiny. After a certain time the surface of the aluminium gets grubby, and the $h_c$ value then starts creeping up to between 3 and 3·5 W/m² K.

If one has a very wide gap, it often pays to suspend aluminium foil loosely in the cavity, thus effectively making two cavities side by side, each being radiation resistant. At its best such a cavity has an $h_c$ value of only 1·15 W/m² K and even after the aluminium has become oxidised and dirty, one can still reckon with an $h_c$ value of 1·5–1·7. Aluminium foil, which does not face a cavity, has no insulation value whatever. Its use is only valid as a vapour barrier.

## Insulation of floors[11]

The earth itself tends to act as a kind of heat-storage container and achieves over a given time the average temperature of the external air. This means that, provided one goes down a distance of more than one metre or so, the soil temperature is not affected unduly by changes in air temperature. Obviously if low temperatures last for a very long time, the temperature of the soil will also go down by a degree or two, and similarly in summer the soil temperature will be higher than in winter, and tend towards the average of day and night temperatures.

In Great Britain, we usually assess the heating period as eight months of the year.

Using degree-day calculations (explained later on) we can evaluate the average winter temperature as being the following in different parts of the country:

| | |
|---|---|
| Southern England | 6·0°C |
| Midlands | 5·5°C |
| Northern England | 5·0°C |
| Glasgow–Edinburgh belt | 4·5°C |
| Aberdeen and high-lying areas throughout Britain | 3·5°C |

We may assume that this will be the average soil temperature at a depth of 1 metre below ground level and lower during the entire winter months.

To calculate the heat losses through the floor, we must employ exactly the same type of calculation as we have used for the calculation of heat losses through walls.

TABLE 11. *k*-values of soils and so on in W/m K

| | |
|---|---|
| Water | 0·60 |
| Snow | 0·06–0·12 |
| Damp loose soil | 0·70 |
| Dry soil | 0·14 |
| Stony normal soil | 0·52 |
| Very damp, tightly packed soil | 1·1–1·4 |
| Hardcore | 0·55 |

The boundary layer heat-transfer coefficient for a floor is less than for a wall. While we have always assumed an internal $h_i$ value for walls of 8·5 W/m$^2$ K, this value is equal to 5·7 W/m$^2$ K when considering the floor.

Let us now evaluate the *U*-value of a floor consisting of 200 mm hardcore, a layer of polyethylene foil (negligible thermal properties) and a layer of concrete screed 100 mm thick. We assume the *k*-value of the concrete to be 1 W/m K.

We also assume that the soil is tightly packed and damp with a $k$-value of 1·2 and that the temperature 1 metre down is the one to be considered.

$$1/U = 1/5{\cdot}7 + 0{\cdot}1/1 + 0{\cdot}2/0{\cdot}55 + 1/1{\cdot}2 = 1{\cdot}47$$
$$\text{or} \quad U = 0{\cdot}68 \text{ W/m}^2 \text{ K}$$

In a room measuring 5 metres × 6 metres, i.e. with a floor area of 30 metres$^2$, the heat losses encountered would be, assuming an internal temperature of 20°C and the climatic conditions of Northern England:

$$30 \times (20 - 5{\cdot}0) \times 0{\cdot}68 = 305 \text{ watt}$$

Insulation of the floor has a very considerable influence upon these heat losses. The following $U$-values have been calculated, when the floor described previously is insulated with the following materials:

| | |
|---|---|
| 25 mm of expanded polystyrene or foamed PVC | $U = 0{\cdot}45$ W/m$^2$ K |
| 50 mm of expanded polystyrene or foamed PVC | $U = 0{\cdot}34$ W/m$^2$ K |
| 100 mm of mineral wool; floor supported by a timber framework | $U = 0{\cdot}265$ W/m$^2$ K |

Heat losses would then be the following:

| | |
|---|---|
| 25 mm of expanded polystyrene | 202 watt |
| 50 mm of expanded polystyrene | 153 watt |
| 100 mm of mineral wool | 118 watt |

In view of the enormous savings in fuel consumption produced by improving the standard of floor insulation, I find it very surprising that it is hardly ever practised in Great Britain. In other countries, floors are always very carefully insulated because, apart from anything else, cold floors are extremely conducive to a general feeling of discomfort. Heat losses through floors with a relatively small area are greater per m$^2$ than through larger areas, due to edge effects.

Floors which are particularly cold are suspended timber floors where the sub-floor space is ventilated to the outside. These have $U$-values of around 3·0, which means that heat losses would amount to 1330 watt under the conditions mentioned above. Naturally, if floors are well carpeted, heat losses through them will be reduced enormously. Wool has a $k$-value of 0·053 W/m K and foam rubber, which is used as an underlay, has a *k*-value of the same order. This means that 20 mm of underlay + carpet are equivalent to about 15 mm of expanded polystyrene.

### Ceiling insulation

Because hot air rises, the area underneath the ceiling is usually a good deal warmer than the room as a whole. Because of this, heat losses upwards are higher than those experienced through walls and floors. This is shown mathematically by the fact that with ceilings, both the $h_i$ value and the $h_o$ value are equal to 11·5 W/m² K. This compares with an $h_i$ value for walls of 8·5 W/m² K and for floors of 5·7 W/m² K. The calculation of the *U*-value for ceilings is similar to that which applies to walls, except that one has to assume higher boundary-layer coefficients. To calculate the heat transfer of a roof, which is exposed to winds, one has to take into consideration the fact that the external boundary-layer heat-transfer coefficient is affected by the wind.

In Table 6, the $h_o$ value is quoted in relationship to the wind speed. These values are applicable for roofs as well, except that they must then be loaded by approximately 35%. This means that while the $h_o$ value for a concrete wall with wind speed of 5 m/s is equal to 42·4 W/m² K, the value which applies to a concrete roof with the same wind speed is equal to 58 W/m² K.

Therefore, a distinction has to be made in the calculation of the heat transfer through a ceiling, where the top is under wind-still conditions, and the heat transfer through a roof (even if the soffit constitutes the ceiling of the room beneath), where the top surface is affected by the wind. A ceiling which consists of plasterboarding only has a *U*-value of about 5 W/m² K, which is a very poor value indeed and causes very rapid heat losses. It is now common to insulate ceilings with about 25 mm of a high void material, and in certain of the newer projects even 50 mm thicknesses are being used. The author has calculated a thickness of 100 mm as being economical in most parts of the British Isles, i.e. at a thickness of that order, a return of capital of some 12% p.a. will be obtained for the last 20 mm thickness of insulation material, provided the room underneath is being kept at 20°C.

## Cold bridges[12]

If the insulation layer is short circuited at a given point or along a given line, the heat loss there is much higher than the value calculated by taking the area of the spot or line into consideration only. The reason for this is that heat does not only flow parallel to the wall, floor or ceiling but also in an oblique direction. Such a spot or line is called a 'cold bridge' and must at all times be avoided in construction, not only because of the high rate of heat loss there, but also because cold bridges cause condensation, pattern staining or both. It can be estimated that in concrete-frame buildings employing highly insulated wall panels the overall *U*-value of the building is worsened by 10% unless the joint between the external wall and the floor is suitably insulated from the outside. Metal wall ties which are employed to hold the inner and outer concrete leaves together in the case of precast concrete sandwich walls worsen the *U*-value of the wall by about 6–8% even if as few as four ties per metre² are employed.

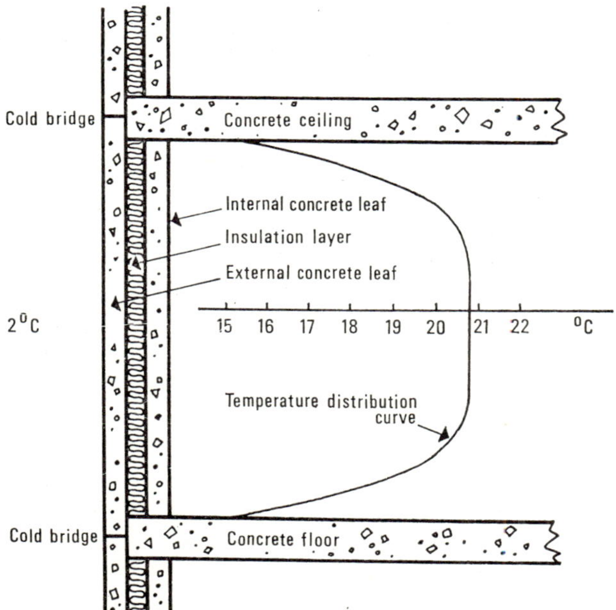


FIG. 7 Cold bridging effects produced in a dwelling built from precast concrete panels

Cold bridging effects are particularly marked with windows. Metal windows where the steel or aluminium goes right through from the inside to the outside of the buildings are rather unsatisfactory. Apart from the fact that heat losses through the metal frames are high, they suffer very badly from condensation and are liable to corrode rapidly. It is far better to construct metal windows in such a way that the two metal surfaces are separated from each other by a layer of badly-conducting material, such as timber, plastic, or particularly a cellular plastic material such as polyurethane or expanded polystyrene.

*Pattern staining*

Cold bridges are the main reason for the phenomenon of pattern staining. If part of a wall or ceiling surface is colder than the surrounding areas, there will be a tendency for dirt to be deposited on this colder area. The reason is that dirt particles are being propelled through the air by impact with air molecules (Brownian movement). Cold air molecules move slower than warm ones, and hence the dirt particles are being driven slowly to the colder areas and deposited on the exact position of the cold bridge. Condensation also naturally centres on this cold spot so that with a plentiful supply of water and dirt there is a good chance for mould formation to start.

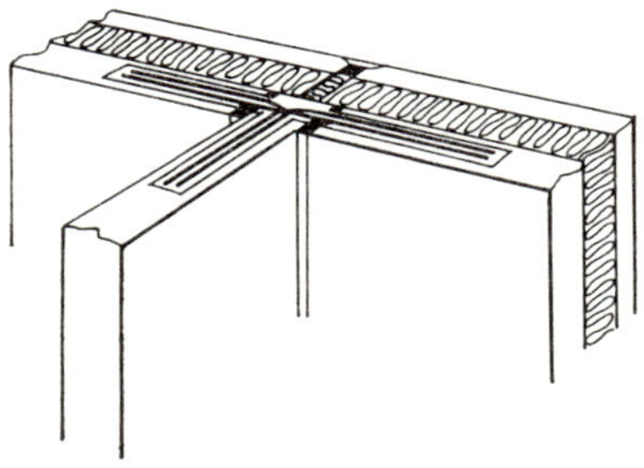

FIG. 8 Wall joints in prefabricated concrete panel structure which obviate cold bridges

It is frequently seen that pattern staining affects the internal corners of rooms rather badly. The reason is that corners of this type constitute a cold bridge unless extra insulation is used. Dirt deposits readily on wallpapers close to metal window frames for exactly the same reason. In the case of uninsulated ceilings the surface is often rather badly stained, but the position of the rafters is traced out in the form of light streaks.

The reason is again that dirt deposits preferentially on the uninsulated ceiling, and is repelled by the surface below the rafter, as this surface is insulated and therefore much warmer. In contrast, it may be noted that warm internal surfaces such as expanded polystyrene ceiling tiles and so on hardly get dirty at all even when the atmosphere is polluted. The reason is that warm air flows down from the surface and keeps the dirt particles suspended, until they eventually pass outside via ventilation gratings. As a general rule one can say: *A well-insulated dwelling is a clean dwelling,* and the reverse is true as well.

FIG. 9 Good insulation with fibreglass quilt or other high void materials prevents pattern staining and also greatly reduces heat losses

**Reflective insulation materials**[13]

Polished aluminium foil is often used as an insulator. In contrast to the high-void insulators (mineral wool, expanded polystyrene, etc.) aluminium foil has no effect on conduction heat transfer, but it reflects some 90% or more of the heat which is transferred by radiation. It can only do this, however, if the aluminium foil faces a layer of air. The proportion of heat which is transferred by radiation varies, however. In the case of walls, 70% of the total heat loss is due to radiation, while in the case of ceilings only 50% is transferred in this way. Therefore the use of aluminium foil is useful if one wishes to cut down heat losses through the ceiling by a moderate amount. One can achieve a reduction in the *U*-value of a ceiling of about 70% by using aluminium foil stretched over the rafters. But when really thick layers of high-void materials are laid, a very much greater reduction in the *U*-value can be achieved. Aluminium foil is far more effective for the insulation of floating floors. In such a case some 70% of the heat losses involved are due to radiation and as aluminium foil can stop these with about 90% efficiency, a layer of aluminium foil pinned to the underside of the floor joists reduces the heat transfer to about 25% of the original value. Again, of course, it is absolutely essential that the aluminium remains clean, that it faces a layer of air and is not embedded in other materials.

## Heat loss through windows

While undoubtedly windows greatly contribute to the provision of heat to a dwelling, even in midwinter, they are also the source of some of the largest heat losses. The reason for this apparent paradox is the following: heat gains through windows are due to radiation. This means the passage of rays of heat energy from the sun, or even from light parts of the sky, through the glass and into the room. Radiant heat rays can pass through a vacuum, and in many ways are similar to light rays, both of which travel at a speed of 300 megametres/second (300 000 km/s).

Heat is lost through windows by conduction and then to the outside air by convection, i.e. much more slowly. Therefore, in winter it happens that heat is constantly being gained through the window by radiation, yet is being equally quickly or even more quickly lost by being conducted to the outside. It is best to consider the two phenomena entirely separately. The heat *gain* through windows facing different compass directions and different localities over the year has been dealt with. Let us now deal with heat *losses* through the windows.

A single-glazed window is basically a very thin lamina of material with a heat resistance which is practically zero. The only resistance to heat transfer is in the two boundary layers, one on the inside and the other on the outside of the pane. The $h_i$ value equals 8·5 W/m² K for vertical windows and 11·5 W/m² K for roof lights which lie horizontally. For inclined roof lights, the value comes out at $(8{\cdot}5 + 3 \sin \theta)$ W/m² K where $\theta$ is the angle the roof makes with the vertical. This means that with a roof sloping at 45°, $h_i = 8{\cdot}5 + 3 \times 0{\cdot}707$ or 10·6 W/m² K.

The external boundary layer heat-transfer coefficient depends on the wind speed and is the following for different types of windows:

TABLE 12. $h_o$ value in W/m² K[14]

| *Type of window* | *Vertical* | *Inclined at 45°* | *Horizontal* |
|---|---|---|---|
| *Wind speed in metres/second* | | | |
| 2 | 12·8 | 16·0 | 17·3 |
| 5 | 26·5 | 32·0 | 35·8 |
| 10 | 46·5 | 58·2 | 63·0 |
| 20 | 81·3 | 100·4 | 110·0 |

Again, of course, the heat-transfer coefficient $h_o$ is related by a sin formula, and $h_o$ can be evaluated for different roof slopes by writing:

$h_o = (12{\cdot}8 + 4{\cdot}5 \sin \theta)$ W/m² K, where $\theta$ is once more the angle between the slope of the roof and the vertical (at 2 m/s).

WORKED EXAMPLE

Determine the $U$-value of a roof light inclined at 45° when the wind blowing past the window has a velocity of 5 m/s. We can write:

$1/U = 1/10{\cdot}6 + 1/32{\cdot}0 = 0{\cdot}126$

Therefore $U = 7{\cdot}95$ W/m² K

*The effect of double glazing a window*[14]

By using an extra pane of glass, an air space is left between the two panes which very much improves the insulating power of the window against conduction heat losses. Radiation heat gain is, however, only reduced by about 10%. The amount of insulation which is provided by a double-glazed window depends markedly upon the following factors:

(a) distance between the panes
(b) strength of external wind
(c) inclination of window to the vertical

FIG. 10 Pilkington's Insulight double glazed units installed in private house in Woking. Architect: Leslie Gooday ARIBA MSIA

The insulating power of an air gap rises almost linearly with the width to reach a maximum at 20 mm width. After that the thermal insulating power remains constant. This means that a double-glazed window with an air gap of 20 mm is as effective a thermal insulator as a window where the gap is, say, 130 mm or 260 mm. But if the air gap is less than 20 mm, its insulating power goes down as a straight fraction. This means that while an air gap of 20 mm has an $h_c$ value of 5·7 W/m² K, an air gap of 3 mm has an $h_c$ value of

$$\frac{5{\cdot}7 \times 20}{3} = 38 \text{ W/m}^2\text{ K}$$

or in general we can say that the $h_c$ value $= 114/t$ W/m² K where $t$ is the thickness of the gap in millimetres.

For windows inclined at 45°, the $h_c$ value for air-gap widths of 20 mm and above equals 7·1 W/m² K and for horizontal roof lights it equals 7·7 W/m² K. The same $h_o$ values as for single glazing apply.

WORKED EXAMPLE

Determine the $U$-value of the same window as before but now double glazed with an air gap of 20 mm.

$1/U = 1/10{\cdot}6 + 1/7{\cdot}1 + 1/32{\cdot}0 = 0{\cdot}267$
Therefore $U = 3{\cdot}75$ W/m² K, a worthwhile improvement over the $U$-value of 7·95 W/m² K obtained previously.

In general, double glazing is at its most effective when it is installed in windows exposed to considerable air currents, where it can reduce heat losses by up to 75%. In the case of rooms which are kept at a low temperature throughout the year, double glazing often shows only marginal economic advantages.

*Types of double glazing*

Double glazing comes on the market in three distinct categories:

(1) hermetically sealed double glazing
(2) coupled double glazing
(3) converted double glazing

*1 Hermetically sealed double glazing*
In these, two panes of glass are assembled in a frame under dry atmospheric conditions, and great care is taken that there is not even a pinhole in the set-up through which moist air can diffuse into the gap between the panes. When these hermetically sealed units first came on to the market the methods of making the seals really tight had not yet been perfected and most of them gave trouble. The author is familiar with a number of

large contracts which were completed 5–10 years ago, and where every single double glazing unit has broken down. In consequence the windows are obscured by condensation between the panes both summer and winter. Unfortunately, once moisture has penetrated the gap between the panes in a hermetically sealed unit, it is never able to escape again.

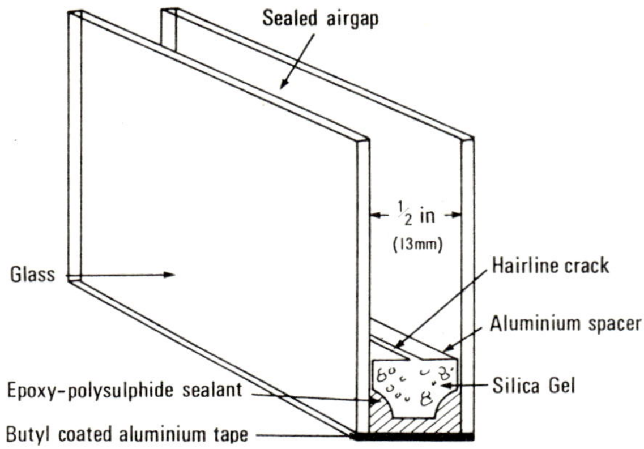


FIG. 11 A section through one edge of a new Rentokil sealed double glazing unit

The trouble does not seem to arise with units which were manufactured more recently, as considerable studies have been made by the major glass firms with a view to getting the hermetic seal foolproof. A very reliable method in this field is employed by Pilkington in the *Glastoglass* units, where the window consists of a single glass bottle, i.e. where there is no reliance upon joints made of metal, rubber, plastic, etc. But one of the disadvantages which still remains with hermetically sealed units, even if they are reliable from the breakdown point of view, is that the air gap is usually well below 20 mm in width, which is the minimum width that is best from the insulation point of view. This can be overcome by using three or more panes in the form of triple and quadruple units so that there are two or three air gaps in the window. Such units are, however, rather expensive.

*2 Coupled double glazing*

These types of double-glazed windows are by far the most popular in Europe, and are satisfactory from every point of view. Basically they consist of two timber, or metal and timber, frames which are hinged to each other. Normally the frames are coupled together by means of hooks, screws or other coupling devices so that the two frames move as one. Glazing of double-glazed windows of this type can be done with the cheapest window glass, and no particular accuracy of manufacture is necessary. Unlike hermetically sealed double-glazed windows, they cannot break down and the air gap can be made as wide as is convenient.

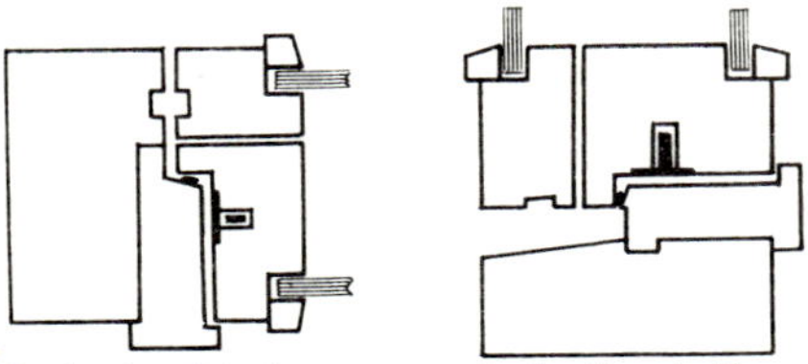


FIG. 12 Sections through a typical coupled frame double-glazed window (Carda). Note position of ventilation ducts

Air is not excluded from the air cavity between the two panes, but care is taken in the design that there is at all times an easier access of air from the outside of the dwelling to the cavity than from the inside. This means that air is stopped from entering the cavity from the inside of the house by the provision of a rubber or foam-plastic seal. On the other hand, the cavity is vented to the outside by means of a crack, roughly 1 mm wide all around the frame, to enable moisture drifting through the timber frame into the cavity to be dissipated to the outside. This 1 mm air gap does not in any way reduce the insulation efficiency of the window. In fact it was shown that its width can be increased to 2 mm before affecting the window in any way. When the glass surfaces on the inside of the air cavity get dirty, the window is uncoupled, and the glass cleaned in the normal way.

Double-glazed windows of this type on the British market today are mostly imported from abroad and because of high customs duties, etc., they are very expensive. Yet from the design point of view they are very easy to make and should be well within the capability of even a moderately sized British window manufacturer. In many countries standardisation and well-developed mass-production techniques have made it possible to make *double*-glazed windows at costs which are not radically different from those applying to *single*-glazed windows in the UK.

*3 Converted double glazing*

Because of the undoubted enormous potential market for double glazing in Britain, numerous systems of converting existing single glazing to double glazing have mushroomed. Unfortunately nearly all of these consist of an extra frame to be attached to the *inside* of the existing window. With such a system it is very difficult to prevent the deposition of condensation and consequent fogging on the inside surface of the external pane.

The water-vapour pressure inside a dwelling is always a good deal higher during the winter months than that on the outside, even on the wettest and rainiest of days. For example, let us assume that the inside humidity is 50% (a very moderate figure, and one which is always exceeded in bedrooms, kitchens and bathrooms) and that the internal temperature is 20°C. The saturation water-vapour pressure at this temperature is equal to 2·34 kN/m² and this makes the actual water-vapour pressure $0{\cdot}5 \times 2{\cdot}34 = 1{\cdot}17$ kN/m². If the external temperature is equal to 5°C and the relative humidity is 100%, as it might well be on a day when it is raining 'cats and dogs', the actual water-vapour pressure is still only 0·87 kN/m², while at 0°C and 100% relative humidity the external water-vapour pressure equals 0·61 kN/m².

It is essential, therefore, that the rate at which water vapour is allowed to diffuse from the air cavity to the outside is always much faster than that at which water can diffuse into the cavity from the inside. This means ventilation towards the outside and hermetic sealing towards the inside. Any imperfections in the sealing of the internal frame to the external one means heavy condensation which is unable to evaporate again. For converted windows of this type to work, they must be very carefully constructed with foolproof sealing to the existing frames, and the timber of the frames must also be flawless. In addition, it is necessary to have large ventilation holes leading from the cavity to the outside to dissipate any moisture which drifts into the cavity from inside the dwelling via the timber.

It is far better to have converted double-glazing systems where the extra frame is hinged loosely to the *outside* of the

FIG. 13 A Crittall double-glazing conversion. Room air is excluded from the cavity by means of sponge rubber seals around the new frame, which clips to the inside of the existing frame. The space between the two panes must also be vented to the outside

existing windows. Provided the top edge is duly protected against the ingress of driven rain, it is impossible for water to penetrate the cavity and obscure the pane. The author fitted such a double-glazing conversion in his own house at a negligible cost seven years ago, and this has shown itself to be absolutely foolproof in all weather conditions. Yet many acquaintances of his, who have fitted traditional types of double-glazing conversions, where the extra frame is fitted on the inside of the existing frame, complain bitterly about the fact that the window seems to be almost permanently obscured by condensation, although the conversion systems were by no means cheap ones.

### Comfort conditions[15,16]

The question often arises as to the optimum temperature inside a dwelling. It is certain that temperatures which are too low are not only undesirable, but also positively dangerous. The normal temperature of the human body is around 37°C. We know that when we are ill, our temperature goes up. But what is not so generally known is that, if people are left in cold surroundings, the body temperature can drop quite markedly and body temperatures as low as 34°C have been recorded in the past. Under such circumstances the metabolism of the human body slows down and the body can no longer defend itself against bacterial invasion. Dr Geoffrey Taylor made a study of so-called hypothermia as he was interested in its effect upon troops, arctic explorers, and so on. He found that, as these were generally young people, there was no marked long-term effect. He found that the people worst affected by hypothermia were old people living in underheated and underinsulated dwellings. He estimated the death rate in England and Wales from diseases associated with hypothermia at between 20 000 and 100 000 per year.

In addition, Dr Taylor[15] studied the efficiency of people working in inadequate temperature conditions and found it much impaired. In offices there seemed to be a considerable difference in working conditions near large single-glazed windows, which induced considerable body-heat loss, and elsewhere.

What should internal room temperatures really be? In nearly all European countries and in the United States, room temperatures varying from 23°C for living rooms to 18°C for bedrooms are usually recommended as a minimum standard. British standards are much lower. The Parker-Morris standard of 1961 and the more recent Ministry of Housing and Local Government report, *Space and Water Heating* (1967)[10] recommend the following standard:

| | |
|---|---|
| Living rooms | 21°C |
| Kitchens, halls and bathrooms | 16°C |
| Bedrooms | 13°C |

How have these figures been obtained? No mention whatever is made in either report of this and one must be forgiven for thinking that the figures were obtained by the simple process of 'think of a number. . . .'

Let us deal with the bedroom temperature. It is possible to sleep in comfort in a bedroom of any temperature, provided one uses enough personal insulation, i.e. enough bed coverings. One can estimate that the average human being loses about one watt per kilogramme of body weight through his or her skin while sleeping. Comfort conditions are only attained when the heat loss is exactly equal to this. If it is in excess of this figure, the body cools down too quickly. Other body functions start to suffer as metabolic processes have to be diverted to the task of keeping the body at a constant temperature in face of these high heat losses. The nervous system reacts quickly by 'feeling the cold'. If the degree of insulation is such that the body cannot get rid of this quantity of heat through the skin, attempts are then made by the body to get rid of it in other ways, i.e. by perspiring and so on. Again discomfort is felt by the nervous system.

The bedroom temperature recommended by the Ministry of Housing and Local Government is too low. The European norm is around 18°C, where full comfort conditions can be achieved with fewer bedclothes. Apart from this, a temperature of 13°C means that there is quite a lot of discomfort when one dresses and undresses. In bedrooms where infants are sleeping one should have a much higher temperature. If the infants kick their bedclothes off, low bedroom temperatures can not only be uncomfortable but also dangerous.

Naturally it is possible for people to sleep with reasonable comfort in a bedroom which is kept at a much lower temperature than this. All that is necessary is more blankets, an eiderdown or two, as well as a few artificial aids to personal heating such as an electric blanket or a hot-water bottle. Many people seem to prefer to sleep in a completely unheated bedroom, with the windows wide open, and use the methods just described to keep their bodies warm. It should, however, be mentioned that on the whole it is the older element of the population which likes to sleep like this because it is not easy to break a habit of a lifetime. Before the advent of central heating, this method of keeping the body temperature up at night was a necessity, not a matter of choice. Naturally no one should be forced to have their bedrooms at 18°C or indeed at any other arbitrary temperature. It should be possible to get bedroom temperatures up to this or even a higher level, particularly if infants are sleeping there or if someone is ill.

16°C for a bathroom is too low. After all, people are usually partly or fully undressed in bathrooms, and in consequence severe discomfort may be experienced. Bathroom temperatures should be higher than living-room temperatures, not lower.

21°C for living rooms is just about right, although it should be possible to get a higher temperature if so desired. The level of temperature in a living room depends largely upon personal preference, which in turn is governed by the kind of clothes

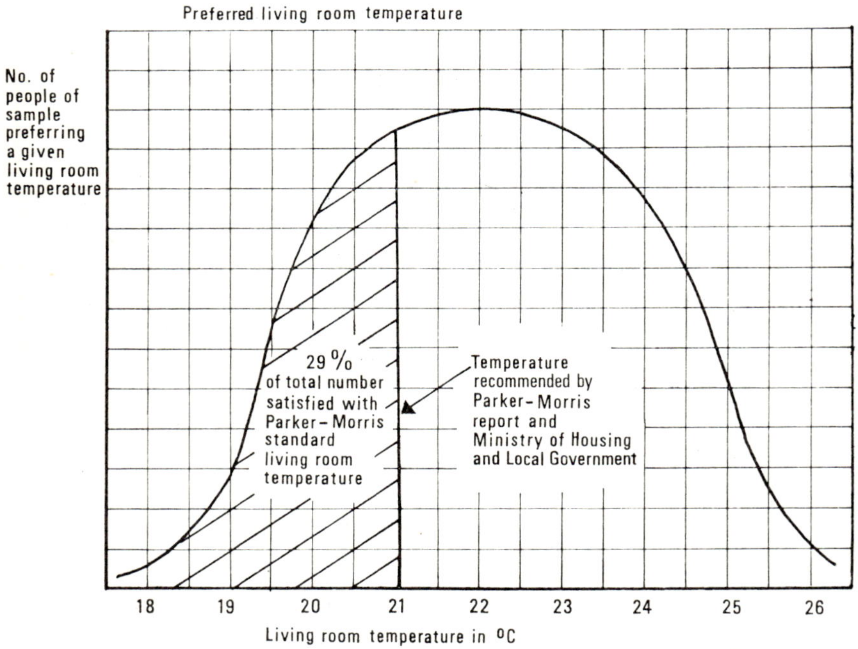


FIG. 14 Graph showing preferred living room temperatures and the comparatively small number of people satisfied with Parker-Morris standard living room temperatures as recommended by the Ministry of Housing and Local Government

people are comfortable in. A recent trend has been, particularly in the case of women, for thinner and lighter clothes to be worn indoors. This means that indoor temperatures have naturally to become higher if one wishes to avoid discomfort. Little research as to the preferred level of indoor temperature in dwellings has, to the best of the author's knowledge, ever been carried out in the United Kingdom. But considerable work has been done in this field both in the United States and in Europe. Figure 14 shows a graph obtained by the American Society of Heating and Ventilating Engineers[25] regarding the level of indoor temperature at which a large sample of subjects feel the most comfortable. As can be seen, the peak is represented by a temperature of 22°C, while there are minority groups who feel comfortable at room temperatures of 19°C and below, while some others do not feel comfortable until the room temperature reaches a level of 25°C or even higher. If the standards of the Ministry of Housing and Local Government are followed, only approximately 29% of the population will ever feel really comfortable in the living rooms concerned. One must therefore design the heating-insulation system to be able to maintain an indoor temperature of up to 25°C in living rooms, halls and bathrooms, and a temperature of around 20°C in bedrooms, but incorporate thermostatic devices to maintain lower temperatures if so required.

Under extreme conditions, particularly of heat, the comfort felt is affected by the relative humidity of the air. If the atmosphere is highly humid, it becomes difficult for the skin to get rid of heat by perspiring. For this reason a higher percentage of the body heat must be given off by normal convection processes.

TABLE 13. Temperatures at different humidities which have the same comfort conditions

| *Relative humidity %* | *Temperature in °C* |
|---|---|
| 40 | 22·5 |
| 50 | 21·9 |
| 60 | 21·3 |
| 70 | 20·7 |
| 80 | 20·1 |
| 90 | 19·5 |

This means that in highly humid conditions a lower air temperature is desirable than in conditions where the air is dry. We do not often get examples of very hot and humid atmospheres in this country. In New York City summers are almost unbearable, not because external temperatures are particularly high, but because the atmosphere is very humid. In contrast, in other parts of the United States much higher temperatures can be borne with equanimity because the air is dry.

**The temperature of internal surfaces in dwellings**

Although, as regards the calculation of the overall heat-transfer coefficient or *U*-value, it does not matter unduly in which order the various building materials are arranged in a wall or in a floor, this is a matter of very considerable importance when comfort conditions are considered. If the insulation materials are placed in the centre or at the outside of the walls, then the internal surface temperature of the wall surface will be comparatively low, and people sitting close to the wall will feel cold because of a radiation heat transfer between their bodies and the cold wall surface. On the other hand, if the layer closest to the inside surface of the wall is highly insulated, this means that the temperature approximates the room temperature and there will be no discomfort. A very good method of doing this in practice is to line the wall with expanded polystyrene slabs, faced on the inside by sheet plastic, timber veneer and so on, to protect the soft polystyrene against physical damage. A number of such walling systems are on the market. These are almost ideal for the construction of the internal walls of offices. Offices, incidentally, must always maintain high internal temperatures, partly because work is sedentary, and partly because of the fact that women tend to wear lighter clothes than men.

In floors it is also essential that the top surface should be made from a highly insulating material; carpets are excellent in this respect. But where it is not feasible to have thick carpets there are other more resilient highly insulating floor coverings which can be used. It is particularly in bathrooms, where people tend to walk around in bare feet, that it becomes essential to have good surface insulation. Where tiles are being used, they should be mounted on slabs of foamed PVC, cork/cement slabs or similar good insulation materials to avoid discomfort. Concrete is very uncomfortable to stand on, even if it is insulated several

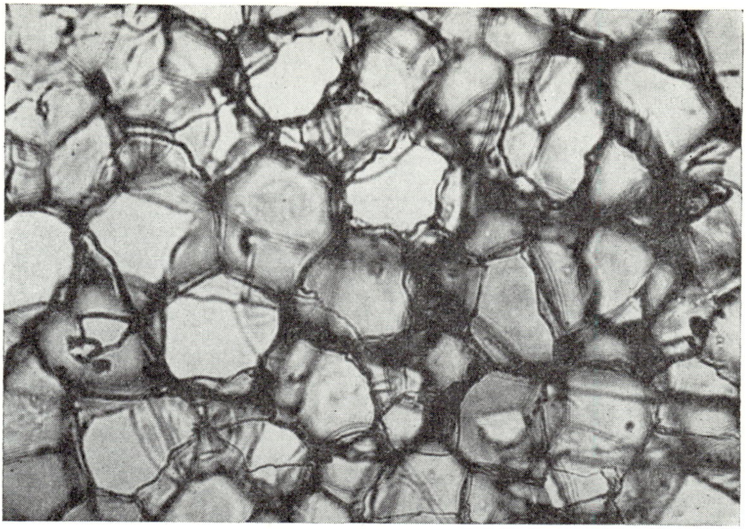

FIG. 15 A microphotograph of expanded polystyrene, showing the tiny bags of air which constitute the insulation layers

centimetres down. Present-day practice of building the ground floors in dwellings by putting thin thermoplastic tiles directly upon the concrete flooring screed is bad. It would not cost much more to cast the floor from concrete, lay a course of thermally insulating material such as gas concrete, chipboard, tongued and grooved boarding, etc. on top, and then follow up with the thermoplastic tiles. Such a floor would be infinitely more comfortable to stand on, either with or without shoes.

## Radiation heating and cooling[17]

The external surface temperature of the walls and roofs of houses and other buildings does not remain constant over the 24-hour period, even assuming that the air temperature remains the same. During the day the surface temperature of the wall is quite a bit higher than the air temperature because the wall surface receives radiant energy from the sun. At night, on the other hand, the surface temperature is lower than that of the external air.

We can write the surface temperature attained by a given

wall under conditions of solar radiation as follows:

$$\text{surface temperature in } ^\circ\text{C} = \text{air temperature} + \frac{I \times a}{h_o}$$

where $I$ is the sum of radiation from sun and sky falling on the wall given in watt/metre$^2$, $a$ is the absorptivity coefficient of the wall surface as given in Table 14, and $h_o$ is the external boundary layer heat-transfer coefficient, which depends upon the strength of the wind blowing past the surface.

TABLE 14. Absorptivity coefficients of different building materials[8]

| | |
|---|---|
| Polished aluminium | 0·04 |
| Rough-surfaced aluminium | 0·06 |
| Heavily oxidised aluminium | 0·25 |
| Copper | 0·28 |
| Aluminium paint | 0·45 |
| Steel | 0·55 |
| Galvanised iron | 0·65 |
| Clay tiles | 0·78 |
| Slate | 0·83 |
| Concrete | 0·85 |
| Wood | 0·90 |
| Grey brick | 0·90 |
| Plastic sheeting | 0·90 |
| Red brick | 0·92 |
| Dirty iron | 0·92 |
| Porcelain | 0·94 |
| Asbestos cement sheeting | 0·96 |
| Black paint | 0·99 |

Table 15 gives the maximum radiation from sun and sky to be expected on walls facing different compass directions and also on flat roofs. These figures are given for different parts of the year, and it should also be noted that the time of the day and the duration of maximum solar radiation naturally also differ with the season. Walls facing east get more radiation than walls facing west because they get their radiation in the morning when the walls are still cold and therefore radiation heat exchange is at a maximum.

TABLE 15. Maximum radiation in watt/metre$^2$ [4]

| *Compass directions* | *December* | *March* | *June* | *September* |
|---|---|---|---|---|
| North | 37 | 79 | 215 | 76 |
| North-east | 37 | 264 | 583 | 230 |
| East | 199 | 644 | 734 | 600 |
| South-east | 503 | 771 | 640 | 739 |
| South | 619 | 770 | 538 | 743 |
| South-west | 437 | 546 | 378 | 527 |
| West | 42 | 86 | 110 | 83 |
| North-west | 37 | 79 | 102 | 76 |
| Flat roof | 101 | 482 | 748 | 465 |

WORKED EXAMPLE

Find the maximum external surface temperature of a brick wall facing west on a winter day (December) when the external temperature is equal to 0°C, and when the air is wind-still.

$$\text{surface temperature} = 0 + \frac{0{\cdot}92 \times 42}{8{\cdot}3}$$
$$= +4{\cdot}5°\text{C}$$

This temperature must then be considered in heat-transfer calculations rather than the normal external temperature of 0°C. On the other hand, at night the surface temperature of the wall is lower than the air temperature. The reason for this is that radiation heat transfer does not take place between the wall and the air surrounding the wall, but between the wall surface and the air of the upper atmosphere. This temperature can be anything between −10°C and −50°C. On very clear nights, the radiation heat transfer between the wall surface and the upper atmosphere is very much higher than on cloudy or foggy nights. At its maximum the surface temperature of external walls can be as much as 5°C lower than the external air although if foggy conditions abound, the external wall temperature even at midnight is almost exactly the same as that of the external air.

What this means in practice is that the heat transfer through walls during the day is somewhat smaller than calculated, while that taking place at night is larger.

## Intermittent heating and cooling[18]

As has just been shown, even if external air temperatures remain constant, which they do not, external wall temperatures tend to fluctuate considerably during day and night. But in addition, internal temperatures are often not kept constant either. In schools, offices and workshops, heating is only carried out during the day and the building is allowed to cool down at night. In homes, too, there is a tendency to switch off the central heating system during the night and to start it off again in the morning. Figure 16 shows a typical graph as to how internal and external wall temperatures may vary in a building over a period of, say, 24 hours.

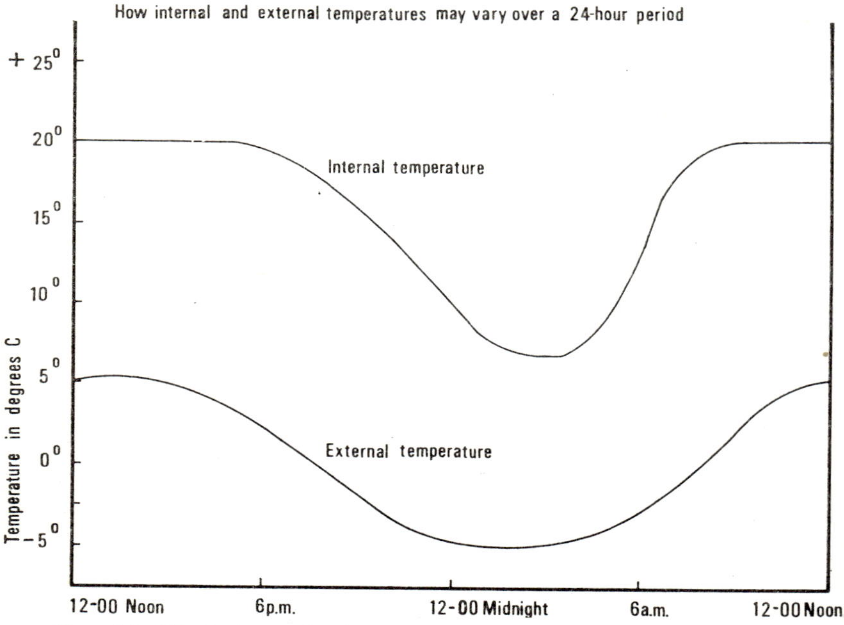


FIG. 16 Typical graph showing how internal and external wall temperatures may vary in a building over a 24-hour period when central heating is switched off

We are often, therefore, not only concerned with the heat-transfer coefficient through the wall, but also with the heat-storage capacity of the wall. Buildings which are constructed from very solid materials, such as concrete bunkers, are capable of maintaining a very steady temperature inside. It takes a long time until the structure is really warm, as heat is abstracted not only to the outside by normal conduction and convection processes, but is also stored in the building shell. When the shell of the building is really warm, the temperature is maintained even if heating is discontinued for some time. Heat losses to the outside come from the heat stored in the building shell rather than from the ambient air on the inside. Similarly, the internal temperature of the building is not markedly affected by a sudden drop of temperature outside or by increased heat transfer due to wind.

Walls of many modern lightweight buildings have a very low heat-storage capacity; for example, walls made of some of the newer curtain wall-cladding materials. If external and internal temperatures remain steady the heat flow through walls with the same *U*-value would be the same, despite the fact that one wall may be a massive structure with a large heat capacity and the other a lightweight curtain wall. But it is a different story when fluctuating temperatures are considered.

When the wall has a low thermal storage, any increase in temperature difference is noted immediately inside the building. In other words, the increase in heat flow that takes place due to the increase of temperature difference between the inside and outside ambient temperatures is entirely at the expense of the room temperature. Unless heating is stepped up immediately the air temperature in the room falls appreciably.

When, however, the wall is of solid construction, the heat which flows to the outside is from heat stored in the body of the wall. The outside layer of the wall is now cooler than it is under steady states, while the warm temperature on the inside is maintained. The governing factor is the speed with which heat can flow from the room into the wall. With very massive constructions this speed is low.

When the temperature difference between the inside and the outside is reduced, the situation is reversed. With a lightweight wall, the heat flow from the inside to the outside reduces almost immediately because the wall has such a low heat-storage effect. But the heavy wall has cooled down considerably during the previous period, and this heat has to be supplied from the inside until steady conditions are once more achieved. We can express all this quantitatively in this way:

The time needed to heat up a cold wall or floor equals:

$$A \frac{t^2 C d}{k} \text{ seconds}$$

where $A$ is a constant depending upon heating conditions, internal and external ambient temperatures, etc.

$t$ is the thickness of the wall or floor in metres

$C$ is the specific heat of the walling material in J/kg K

$d$ is the density of the walling material in kg/m³

$k$ is the thermal conductivity of the walling material in W/m K

For any given wall, the heating-up period varies with the square of the thickness of the wall. It therefore takes four times as long to heat up a concrete slab 200 mm thick than it takes to heat up a concrete slab 100 mm thick, provided all other conditions are the same.

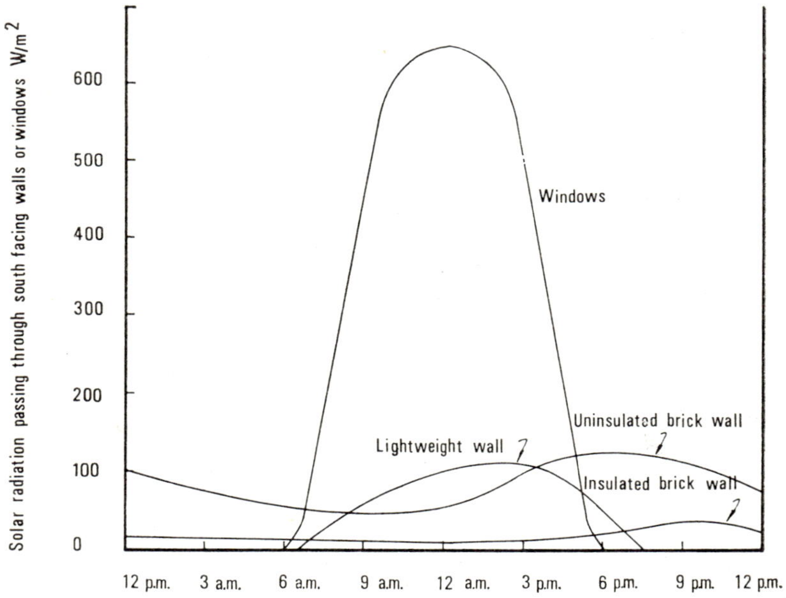


FIG. 17 Passage of solar radiation through three types of wall and windows. Note how heavy construction contributes to even temperatures within the building

To compare walls which are made from different materials we make use of the data in Table 16.

TABLE 16. Thermal data for a number of building materials

| | $k$ (W/m K) | $d$ (kg/m$^3$) | $C$ (J/kg K) |
|---|---|---|---|
| Aluminium | 198 | 2700 | 0·85 |
| Iron | 48 | 7860 | 0·50 |
| Glass | 0·6 | 2600 | 0·50 |
| Granite (and most other similar solid rocks) | about 2·0 | 3000 to 4000 | about 1·0 |
| Brick | 0·7 | 1700 | 0·95 |
| Aerated concrete, vermiculite concrete, porous gypsum, etc. | about 0·1 | about 500 | about 0·8 |
| Water | 0·58 (ignoring convection) | 1000 | 4·2 |
| Expanded polystyrene, PVC, polyurethane, etc. | 0·035 | about 30 | 1·6 |
| Most concrete products | 0·7–1·9 | varies between 1000 and 2800 | 0·85 |
| Wood, chipboard, hardboard, etc. | 0·2 | varies between 400 and 950 | 1·3–1·6 |

Let us consider an aerated concrete wall and a brick wall 0·2 m thick. The aerated concrete wall takes:

$$\frac{0{\cdot}2^2 \times 0{\cdot}8 \times 500}{0{\cdot}1} A = 160\ A \text{ seconds to heat up}$$

The brick wall takes:

$$\frac{0{\cdot}2^2 \times 0{\cdot}95 \times 1700}{0{\cdot}7} A = 92{\cdot}3\ A \text{ seconds to heat up}$$

## When to use heavyweight and when to use lightweight construction

One must at all times use building materials with as low a $U$-value as possible. But the heat-storage capacity and the time needed for a wall or floor to heat up and to cool down under fluctuating ambient temperature conditions is equally important. For buildings which are occupied all day, such as dwellings, hospitals, old-age pensioners' homes and so on, one should always use building materials where the thermal storage effect is high, i.e. where the period of warm-up and cool-down is as long as possible. In the opinion of the author, lightweight curtain walling is totally unsuited for such purposes. It can only be justified if insulation values several times as high as would be reasonable with heavy construction are used.

Fig. 18 Aluminium foil used in conjunction with a high void insulator

Lightweight curtain wall materials should be used for buildings which are being used for a few hours per day only, such as schools, offices, workshops and similar buildings. The low thermal storage capacity of the walls means that such buildings are heated up quickly in the morning, and that little heat is wasted in replacing heat that has been lost from the building at night. Under such circumstances one should use walls and floors which are so constructed that the time of heating up (and cooling down) is minimum. During the few hours per day that heating is required, it is unlikely that radical external temperature changes will take place and therefore the disadvantage of a poor storage capacity is not an overwhelming one. Admittedly, differences in wind speed will make themselves felt and therefore buildings of this kind should have heating systems which can adjust themselves quickly to varying heat demands.

Once the day-time use has finished, all heaters are turned off and within a few hours the building cools down virtually completely. Because of the low storage capacity of the walls, the quantity of heat lost in this way is not large. Lightweight construction is thus suited to the construction of schools, day nurseries, offices, workshops and so on, but not for homes and hospitals.

## The degree-day concept[19]

It is necessary to evaluate just how cold a given area of the country is to be able to work out whether a given degree of insulation is likely to be profitable or not. This is done most conveniently by working out the annual number of degree-days. Let us assume an arbitrary temperature, which we take as the figure giving the average daily heat level in a house. This is commonly taken as 18·3°C. If, on the day in question, the average external temperature is equal to 4°C this means that on that day there are $18{\cdot}3 - 4 = 14{\cdot}3$ degree-days. The annual number of degree-days constitutes the sum of these degree-days over the entire 365 days of the year. Naturally, one discounts negative degree-days, i.e. when the external temperature is higher than 18·3°C.

In Great Britain, at sea level, the annual number of degree-days is mainly governed by latitude. The following values apply to the main industrial belts in Great Britain:

| | |
|---|---|
| London–Bristol | 2940 degree-days |
| Birmingham | 3050 degree-days |
| Manchester–Sheffield | 3200 degree-days |
| Newcastle–Carlisle | 3320 degree-days |
| Glasgow–Edinburgh | 3470 degree-days |
| Aberdeen | 3740 degree-days |

These values apply to fairly rural areas near these big cities. In urban areas, the annual number of degree-days is reduced by about 300 degree-days per annum. Degree-days increase rapidly as the altitude increases. Each 100 metre altitude above sea

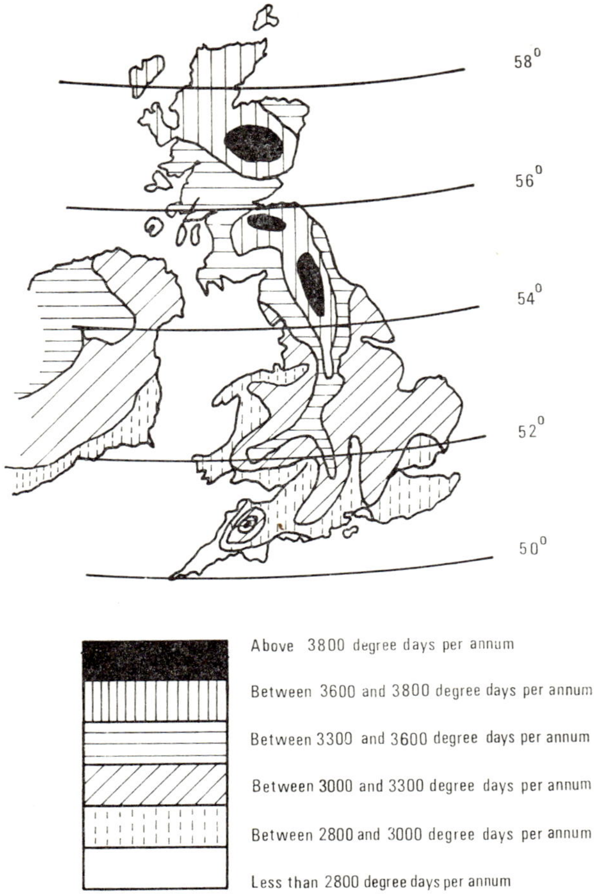


FIG. 19 Approximate number of degree-days in the British Isles using the Centigrade system and base temperature of 18·3°C

level adds on average 100 degree-days to the overall figure. For example, high-lying southern areas like Dartmoor have virtually the same number of annual degree-days as places in Northern Scotland which lie at sea level.

Although the annual number of degree-days in Great Britain appears to be mainly dependent upon the latitude and altitude, there is a considerable distinction as to the actual duration of the heating season and the lowest temperatures experienced. Places on the eastern side of Great Britain tend to have a more extreme climate. This means that they have colder and shorter winters, but also longer and warmer summers. On the east coast of Great Britain one assumes a heating season of about 7½ months in the year. On the western side of the country, winter temperatures do not fall as low, but the winter lasts longer. There we assume a heating season of about 8½ months in the year. The average heating season therefore varies between 255 days and 225 days per annum, depending upon whether the place in question lies to the west or east of the country.

## The economics of insulation [20]

The annual cost of heating and insulating a given area of wall is given by the following equation:

$$C_a = \frac{p\,B}{100} + 0{\cdot}0864\ U\ Y\ a + C_s + C_m$$

£ per metre² of wall

where $p$ equals the sum of the % interest and depreciation of the capital cost of insulating the wall, as given in Table 17

$B$ is the capital cost of putting in the insulation in £/m² of the wall

$Y$ equals the annual degree-day value of the site

TABLE 17. Capital repayment and interest charges for different write-off periods and interest rates

| *Years of depreciation* | *Interest rate* % | *Total rate* % | *Interest rate* % | *Total rate* % |
|---|---|---|---|---|
| 10 | 7 | 17 | 8 | 18 |
| 20 | 7 | 12 | 8 | 13 |
| 30 | 7 | 10⅓ | 8 | 11⅓ |
| 40 | 7 | 9½ | 8 | 10½ |
| 50 | 7 | 9 | 8 | 10 |
| 10 | 9 | 18 | 10 | 20 |
| 20 | 9 | 14 | 10 | 15 |
| 30 | 9 | 12⅓ | 10 | 13⅓ |
| 40 | 9 | 11½ | 10 | 12½ |
| 50 | 9 | 11 | 10 | 12 |

$U$ equals the $U$-value achieved

$a$ equals the cost of heating in £ per useful megajoule

$C_s$ gives the additional cost involved due to the reduction in space by virtue of putting insulation on, in £/m$^2$ of wall (This is zero when the insulation is being put into a cavity, which is there anyway, but naturally enters when the cavity is made bigger on purpose.)

$C_m$ is the extra cost of maintaining the insulation layer per annum, again given in £/m$^2$

If we now write down the following equation, we can read off directly what the most economic $U$-value of a wall should be. It is simply:

$$U_e = \sqrt{\frac{k\,p\,C_{tot}}{8{\cdot}64\,a\,Y}}$$

where $U_e$ is the economic $U$-value in W/m$^2$ K of the wall

$k$ is the $k$-value of the insulation material in W/m K

$p$ is the sum of the % depreciation and interest

$C_{tot}$ is the total capital cost of insulation including the charges to be made for maintenance and loss of space over the whole period in £/m$^3$

$a$ is the cost of heating as £ per useful megajoule

$Y$ is the degree-day value for the area in question.

This equation, however, only works when the building is to be heated for 24 hours per day to an average temperature of 18·3°C.

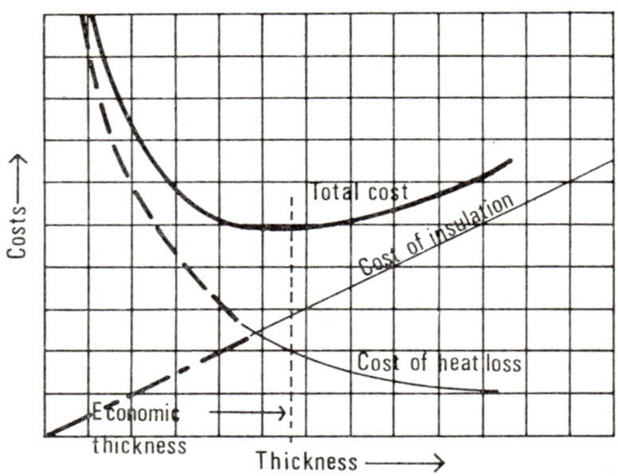


FIG. 20 Graph illustrating the economic criteria of thermal insulation

If a different internal temperature is employed, the degree-day value must be loaded by 240 for each degree C above 18·3°C and by −240 for each degree C below 18·3°C. When buildings are only heated for part of the day, the degree-day value must be reduced drastically.

If we consider a lightweight building which is occupied for 8 hours per day only, then we assume a degree-day value half of the usual amount. For heavyweight buildings, the factor by which we multiply the degree-day value under such circumstances is only 0·85. In other words, one only saves about 15% of one's heating costs by letting the building cool down after 8 hours' occupation.

These values are not accurate but purely approximate, as they depend on the exact nature of the building. They will, however, serve to enable an approximate calculation of the degree of insulation to be aimed at.

WORKED EXAMPLE

Determine the economic $U$-value for a walling system in a building which is being heated night and day to an average indoor temperature of 20·3°C. The building is situated at a height of 200 metres above sea level in central Scotland, and expanded polystyrene is being used as an insulation material. This has a $k$-value of 0·035 W/m K and costs £15 per metre$^3$ to fix in material, labour and all other costs. The heating costs are assessed as £1·25 per 1000 megajoules of useful heat. The insulation system has to be written off in 40 years and interest charges on the capital are 9%.

$Y$ equals $3470+300+480=4250$, $p$ equals 11·5 and $a$ equals 0·01125

Substituting in our equation we get

$$U_e=\sqrt{\frac{0{\cdot}035\times 11{\cdot}5\times 15}{8{\cdot}64\times 0{\cdot}00125\times 4250}}\ \text{W/m}^2\ \text{K}$$
$$=0{\cdot}365\ \text{W/m}^2\ \text{K}$$

How are we going to work out what thickness of expanded polystyrene this represents?

We use exactly the same reciprocal equation which we have been using all along. If we know that the $U$-value of the wall without insulation equals $U_o$, then the thickness of insulation with an insulation material with $k$-value $K$ will be equal to $t$ metres. We can therefore write:

$$1/U_e = 1/U_o + t/k$$

In our case, let us consider that the uninsulated wall is of brick, 230 mm thick, and the uninsulated $U$-value equals 2·3 W/m$^2$ K

Substituting the value we have just obtained we can write:

$$1/0{\cdot}365 = 1/2{\cdot}3 + t/0{\cdot}035$$
$$\text{or } t = 0{\cdot}081 \text{ m or } 81 \text{ mm}$$

This same calculation can be carried out with respect to roof insulation as well. Because roof insulation costs far less than wall insulation, the economics should be even better. A calculation regarding the optimum thickness of roof insulation using mineral wool or fibreglass gives an optimum thickness of such insulation materials as being in excess of 100 millimetres.

It is not suggested that the values which have been used in the sample calculation are valid under all circumstances. They only apply for the data given. But the calculation is so simple that every architect and planning engineer should use it to check whether he is in fact using the most economic thickness of insulation. The author feels that many people will get a shock when they work out how much insulation they should have and how little is actually being used.

*Practical methods for insulating walls*

Many walls of existing buildings were erected in the form of cavity walls. A number or proprietary methods are available to inject the following highly porous materials into these cavities:

(1) foamed ureaformaldehyde
(2) foamed polyurethane
(3) expanded polystyrene pellets
(4) mineral wool

Where buildings are erected from scratch one can use either expanded polystyrene slabs or slabs of mineral wool or glass wool, while existing buildings are best insulated by using expanded polystyrene slabs; these are stuck to the inside wall surfaces and then either plastered or covered with other internal surfacing materials.

# 4 Dampness Inside Dwellings

Damp walls inside dwellings can be caused in two ways:

(1) by the penetration of water from outside the building
(2) by the condensation of water vapour on inside surface

## Driven rain[14]

Rainwater can penetrate the building façade or drip through the roof by certain specific agencies. If there are vertical or sloping gaps downwards, water will penetrate, especially if there is a way for it to lodge above the crack. Also water can be pushed through horizontal gaps, but only if these exceed 4 mm in diameter. Finally, rainwater can penetrate walls by the agency of capillarity. Capillarity is the force which makes tea wet a cube of sugar when it is dipped into it, and is caused by the fact that certain materials are hydrophilic, i.e. they attract water. If a hydrophilic material is also very porous, water rises in the pores. In the case of building materials, all the common brick, stone and concrete materials are very hydrophilic. But there are hydrophobic (water-repelling) materials as well. These are greases, waxes and most plastics. Water in contact with such materials is not attracted by capillary action, but is in fact repelled.

Also most modern building materials are well consolidated and give a good resistance to rain. Mortar and cement are even better in this respect than well-fired brick. Most building materials can only be penetrated by driven rain in the long run, if there is absolutely continuous strong driven rain for many days. This is rare even in Britain. There are

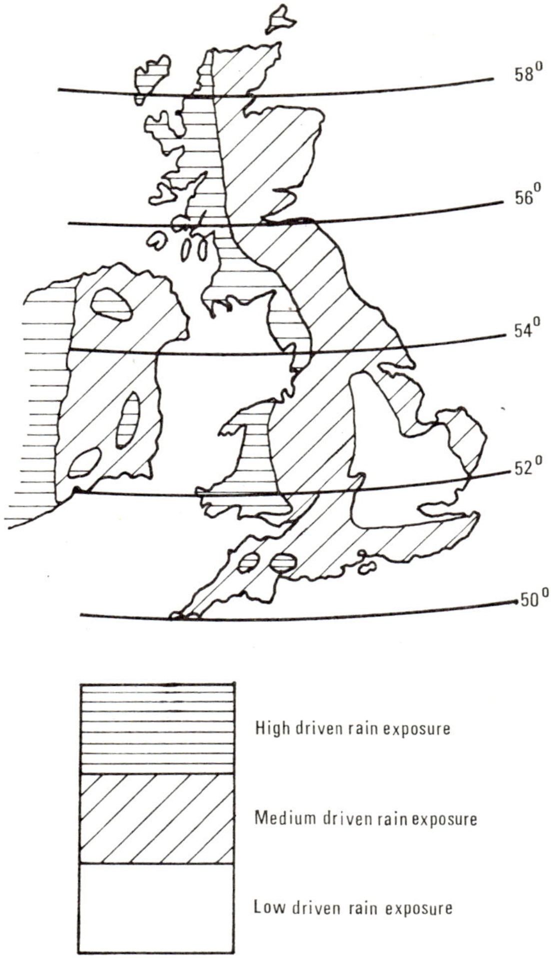


FIG. 21 Map illustrating varying degrees of driven rain exposure in the British Isles

pauses during which the water, which has partially penetrated the wall, can dry out again.

*Prevention of driven rain*

The penetration of a wall by driven rain can be prevented by the following methods:

(1) By leaving a cavity between the internal and external wall leaves, which prevents water bridging over. Care must,

however, be taken that water is not allowed to flow along sloping wall ties or pieces of mortar or other substances which were dropped into the cavity during construction.

(2) By protecting the external surface of the wall with tiles, metal sheets, shiplap boarding or glazed panels and so on. These are always fixed in a loose fashion to permit water vapour to be dissipated from the inside. They are often mounted on top of sheets of water-repellent felt.

(3) By including somewhere in the wall a layer of dense concrete, which has a very good water-stopping power because it contains relatively few pores.

(4) The inclusion of plastic foam in the wall stops water from passing through by capillary action, as the foam has a negative capillarity, i.e. it is greasy towards water. Most grades of mineral wool and glass wool, which are coated on the surface with water-repellant substances such as grease, act in the same way.

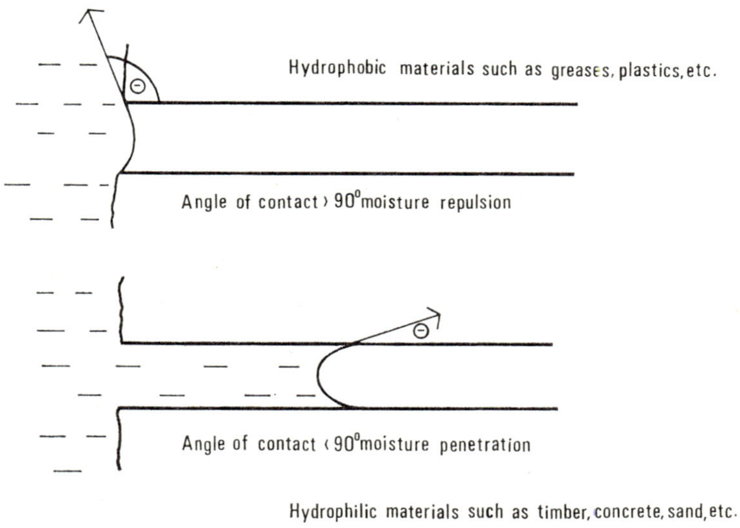


FIG. 22 Diagram showing the influence of the angle of contact upon moisture penetrating through walls

Painting the walls with either silicone or siliconate solution is not included because this is not a particularly satisfactory method.

## Capillary rise of ground water

This is a real danger where there is either no damp-proof course or where the damp-proof course is defective. It is a fault which is unlikely to affect modern dwellings, as they use substantial polyethylene foil sheet underneath the concrete floor slabs which meets modern damp-proof course materials at the wall level. In old houses, a defective DPC or none at all is always a problem. There are many methods available to remedy this defect. There are systems of inserting a whole new DPC by cutting between bricks, as recommended by the British Research Station; there is the Rentokil method of osmotic damp proofing; and there are a number of other techniques around which rely mainly upon bleeding rubber solutions into the pores of the brickwork to produce a water barrier. Silicones and siliconates are often also used for such purposes.

## Condensation

Although there are cases where defective flashing or poor workmanship is the cause of water leakage into the home, by far the most important cause of dampness is not water penetration from the outside, but condensation against walls, windows, ceilings, etc. from the inside.

Now why does condensation take place? Air has a certain carrying capacity for water vapour, which varies enormously with its temperature. This carrying capacity is expressed in newtons/metre$^2$, which is the vapour pressure of the water at the temperature in question.

TABLE 18 Water-vapour pressure of saturated air at different temperatures, and the weight in grammes of water which is carried by one metre$^3$ of the air at this temperature[21]

| *Temperature in °C* | *Water-vapour pressure in N/m$^2$* | *Weight of water carried in g/m$^3$* |
|---|---|---|
| −10 | 286 | 2·35 |
| − 9 | 309 | 2·53 |
| − 8 | 336 | 2·74 |
| − 7 | 361 | 2·91 |
| − 6 | 390 | 3·14 |
| − 5 | 421 | 3·38 |
| − 4 | 454 | 3·62 |

| *Temperature in °C* | *Water-vapour pressure in N/m²* | *Weight of water carried in g/m³* |
|---|---|---|
| − 3 | 488 | 3·89 |
| − 2 | 526 | 4·19 |
| − 1 | 567 | 4·50 |
| 0 | 610 | 4·83 |
| 1 | 655 | 5·17 |
| 2 | 703 | 5·55 |
| 3 | 755 | 5·94 |
| 4 | 812 | 6·40 |
| 5 | 870 | 6·83 |
| 6 | 932 | 7·31 |
| 7 | 1000 | 7·88 |
| 8 | 1070 | 8·32 |
| 9 | 1132 | 8·81 |
| 10 | 1228 | 9·45 |
| 11 | 1310 | 10·08 |
| 12 | 1400 | 11·80 |
| 13 | 1496 | 11·50 |
| 14 | 1590 | 12·20 |
| 15 | 1695 | 13·00 |
| 16 | 1810 | 13·90 |
| 17 | 1930 | 14·88 |
| 18 | 2055 | 16·15 |
| 19 | 2185 | 17·00 |
| 20 | 2330 | 17·90 |
| 21 | 2470 | 18·80 |
| 22 | 2620 | 19·75 |
| 23 | 2790 | 20·75 |
| 24 | 2980 | 21·80 |
| 25 | 3160 | 22·90 |

If warm air, perfectly capable of carrying a given amount of water vapour, hits a cool surface, it will also be cooled. At the lower temperature the air is no longer capable of carrying the amount of water it was able to do before, and the excess is therefore precipitated as condensed liquid water. Air which carries the maximum amount of water at a given temperature, is considered to be saturated or to have a *relative humidity* of 100%.

If we consider air which is at 22°C and which carries 14·8 grammes of water vapour per metre$^3$, this will mean that the air is

$$\frac{14\cdot8 \times 100}{19\cdot75}\% = 75\% \text{ saturated,}$$

or has 75% relative humidity (R.H.)

The dew point is given as the temperature at which the air ceases to be able to carry all this moisture. Looking at our table, we find that the dew point for air containing 14·80 grammes of water vapour per metre$^3$ is just below 17°C. This means that condensation will take place on all surfaces which have a temperature below about 16·8°C. If the air is cooled to 6°C, it will shed about half the water carried by it, while if it is cooled to −5°C it will shed about three quarters of its water as condensation.

Therefore how does one avoid condensation?

*One simply makes sure that air which carries water vapour is never cooled below its dew point.*

## Sources of water vapour

The weight of water which is released in the average occupied dwelling is quite staggering. The main sources are:

(1) water vapour breathed out by the occupants
(2) water vapour released by combustion appliances such as a kitchen stove, paraffin heaters without flue, and so on.
(3) water vapour released by boiling kettles and pots in a kitchen, and by a washing machine when operated
(4) water vapour released by standing areas of water

*Water vapour breathed out by occupants*

It can be calculated that an adult person at rest releases on average 700 grammes of water vapour per 24 hours in the form of perspiration, and also that he or she breathes out quantities of air, varying between 420 litres when at rest or sleeping, and up to 4000 litres when engaged in really violent exercise. If one considers that this air is at 100% humidity and that the temperature at which it is breathed out is 37°C, corresponding to a water-vapour pressure of 6250 N/m$^2$, one can calculate that a human being breathes out 18·4 g of water vapour per hour when sleeping and up to 175 g of water vapour when engaged in really vigorous exercise.

Adding the amount of water released through perspiration at night, which is around 29 g per hour, a person who is sleeping gives out roughly 50 g of water vapour each hour.

*Water vapour released by combustion appliances*

Any hydrocarbon fuel, when burnt, is changed to carbon dioxide and water vapour. We can say, on average, that each kilogramme of paraffin, when burnt, gives 3 kg of carbon dioxide and 1·3 kg of water vapour. Each cubic metre of town's gas, when burnt, gives 700 g of water vapour. Gas fires usually discharge into flues, but water vapour from the kitchen stove is discharged directly into the dwelling unless a fume hood is fitted.

*Water vapour released by boiling water*

An enormous amount of water vapour is released when water is boiled in kettles or pots. It is difficult to evaluate this exactly, but according to an experiment carried out in the author's home, the quantity of water vapour released when a kettle is being boiled, with the gas throttled down to 25% of its normal supply rate, is 640 g per hour. The average washing machine uses a 3 kW heater. Assuming that 80% of the electric power is used to boil the water and 20% is dissipated to heat up the air, the amount of water released during the operation of such a washing-machine boiler is equal to 3600 g of water vapour per hour.

*Water vapour released by standing areas of water*

It is possible to calculate the amount of water evaporating from a given surface by the use of the following equation:

weight water evaporated per hour

$$= \frac{0{\cdot}318 \times A \times K_t \times (P_w - P_a)}{P_a} \text{ g}$$

where $A$ is the area of the water surface in $m^2$

$K_t$ is the temperature constant in $N/m^2 {}^\circ C$ given in Table 19

$P_w$ is the water-vapour pressure of the standing water. This depends on the temperature of the water and is given in Table 19

$P_a$ is the water-vapour pressure of the air. This is given by:

$$\frac{\text{saturation-water-vapour pressure} \times \% \text{ relative humidity}}{100}$$

Both $P_w$ and $P_a$ are given in SI units as $N/m^2$.

TABLE 19. $P_w$ and $K_t$ of standing water at different temperatures[14]

| *Temperature in °C* | $P_w$ *in* $N/m^2$ | $K_t$ *in* $N/m^2{}^\circ C$ |
|---|---|---|
| 5 | 870 | 65 |
| 10 | 1228 | 93 |
| 15 | 1695 | 110 |
| 20 | 2330 | 148 |
| 25 | 3160 | 180 |
| 30 | 4245 | 225 |
| 35 | 5600 | 310 |
| 40 | 7400 | 390 |
| 45 | 9580 | 450 |
| 50 | 12350 | 600 |
| 55 | 15710 | 730 |
| 60 | 19900 | 900 |
| 65 | 25000 | 1200 |

WORKED EXAMPLE

What is the rate of evaporation from a bath measuring 1·2 metres × 0·5 metres if it is filled with water at 45°C. The air temperature in the bathroom is 18°C and the relative humidity is 50%

From Table 19 we find that $K_t = 450$ $N/m^2{}^\circ C$ and $P_w = 9580$ $N/m^2$

From Table 18 we find that the saturation water-vapour pressure at 18°C equals 2055 $N/m^2$, which at 50% relative humidity comes to 1027·5 $N/m^2 = P_a$

Finally we evaluate the area of the bath as 0·6 $m^2$

Substituting in our equation, we can calculate that the hourly evaporation of water equals:

$$\frac{0{\cdot}318 \times 0{\cdot}6 \times 450 \times (9580 - 1027{\cdot}5)}{1027{\cdot}5} = 722 \text{ g}$$

The large quantities of water vapour which are released in an

occupied dwelling must be expelled to the outside if one wishes to avoid condensation troubles.
Water vapour can be lost in two ways:

(1) by diffusion through walls, ceilings, etc. In this case only the water disappears and the air remains
(2) by ventilation, when highly humid internal air is exchanged for dryer outside air

## Water vapour diffusion

The rate at which moisture travels through a wall or ceiling depends on the following factors:

(1) nature of the material
(2) thickness of the material
(3) difference in water-vapour pressure between the inside and the outside
(4) area across which diffusion takes place

The rate of diffusion also varies somewhat with temperature, but under the conditions considered with the average dwelling, errors induced by assuming a standard average temperature between the inside and the outside of 15°C will be small.

The amount of water vapour which diffuses to the outside equals:

$$\frac{6{\cdot}18 \times 10^{-3}\, A\,(p_i - p_o)}{f\,t} \text{ grammes water/hour}$$

where $A$ is the area of the wall in $m^2$
$p_o$ is the external water-vapour pressure in $N/m^2$
$p_i$ is the internal water-vapour pressure in $N/m^2$
$f$ is the diffusion coefficient of the material in question, relative to diffusion through air. This is given in Table 20
$t$ is the thickness of the wall in metres

WORKED EXAMPLE

Give the hourly quantity of water which diffuses through a 200 mm façade brick wall if the internal temperature equals 20°C, with a relative humidity of 60%, and if the external temperature equals 2°C with a relative humidity of 90%. The wall measures 3 m × 6 m.

$A = 18\ \text{m}^2 \quad p_i = 0{\cdot}6 \times 2330 = 1440\ \text{N/m}^2$
$p_o = 0{\cdot}9 \times 703 = 635\ \text{N/m}^2 \quad f = 10$ and $t = 0{\cdot}2$ metres

We can therefore write:
$G$ = hourly diffusion rate =

$$\frac{6{\cdot}18 \times 10^{-3} \times 18 \times (1440-635)}{10 \times 0{\cdot}2}$$

= 44 grammes of water per hour

TABLE 20. Relative diffusion coefficients of various materials[22]

| | | | | | |
|---|---|---|---|---|---|
| Air | 1 | Asbestos cement | 70 | Bituminous vapour barrier | 180 000 |
| Common brick | 5 | Façade brick | 10 | Cellular PVC | 760 |
| Lightweight concrete | 5 | Medium grain concrete | 15 | Heavy concrete | 30 |
| Expanded polystyrene | 45 | Fibreboard | 10 | Glass wool | 1·3 |
| Hardwood | 70 | Rubber | 20 000 | Neoprene | 65 000 |
| Pine | 20 | Plaster | 10 | Roofing felt | 1 300 |

When the wall consists of a series of layers, we use a reciprocal formula similar to the one employed when heat transfers through built-up walls are calculated.

This means that we work out the combined $f\,t$ value by writing:

$$\frac{1}{f\,t} = \frac{1}{f_1\,t_1} + \frac{1}{f_2\,t_2} + \frac{1}{f_3\,t_3} \ldots \text{etc.}$$

where $f_1$, $f_2$ and $f_3$ are the diffusion coefficients of the various wall layers and $t_1$, $t_2$ and $t_3$ are the thicknesses of the wall layers in metres.

*Vapour barriers*[23]

It is essential to avoid the condensation of water vapour inside the walls, particularly if a modern frame wall construction is employed. To achieve this one places a layer of material with a very high $f$-value on the inside wall, and in this way the amount of water vapour which drifts into the wall is very much reduced. If, in addition, the cavity inside the wall is in good open contact with the outside air, there should be no condensation inside the wall. One must, however, emphasise that there is no such thing as a true vapour *barrier*, only a vapour *retardant*. Also water vapour will readily drift through, unless the joints are perfect. Particular attention must be paid to the continuity of vapour barriers at switches, corners, and so forth.

FIG. 23 Injecting ureaformaldehyde foam into cavity walls

As an example of what can happen if the business of vapour barrier is not properly understood, the author would like to quote a case in which he was recently called as a consultant.

Roofs of certain prefabricated houses consisted of a ceiling made from butt-jointed aluminium-foil backed plasterboarding. The external weathering consisted of sheet aluminium, without any degree of ventilation. The result was that moisture from the bathroom underneath was able to diffuse into the cavity, but once inside the cavity it could not diffuse to the outside.

Heavy condensation occurred against the aluminium sheeting; the water dropped downwards and stained the ceiling.

The phenomenon, which can occur in sidings as well, can always be prevented by suitably venting the space on the inside of the wall to the external air. If the rate of exit of water vapour from the cavity is appreciably more rapid than its entry into the cavity via the vapour barrier, then there should be little trouble except under very low temperature conditions. By applying some insulation to the inside surface of the external weathering, the surface temperature of the external siding is raised sufficiently to make condensation far less likely.

FIG. 24 When aluminium sheeting is used as roofing, as in this contract in Northumberland, it is necessary to insulate the underside of the aluminium, and also to provide ventilation of the cavity underneath, otherwise heavy condensation may take place

## Dissipation of water vapour by ventilation

Diffusion through the walls of kitchens and bathrooms is normally totally inadequate to get rid of more than a fraction of the amount of water vapour produced. In bedrooms, the problem is not quite as serious. Each person gives out approximately 50 g of water vapour per hour when sleeping. Provided that the bedroom is kept at a reasonably warm temperature, natural diffusion through walls and ceiling should be able to dissipate nearly all this water vapour. The rest will probably drift through the house, and finally disappear through imperfections at windows and doors. But in the average bathroom as much as 1000 g of water vapour per hour can be released at certain times of the day, and in kitchens figures of 3500–4500 g of water vapour per hour are not uncommon. Natural diffusion cannot eliminate more than a fraction of this quantity. The only answer is ventilation. Naturally, one can always leave the windows open, but this is hardly the best way in winter because one loses an enormous amount of heat as well.

The amount of ventilation which must be practised depends upon the degree of insulation used in the dwelling. This means one must use sufficient ventilation to reduce the water-vapour content of the air to a level when the dew point is no longer reached against the coldest surface in the house. Very highly insulated dwellings with double glazing, effective wall and window insulation, etc. need far lower rates of ventilation than places where insulation standards are crude.

Often, the opposite seems to happen. Places that are inadequately insulated are badly ventilated as well, in order to conserve heat, so that heavy condensation is virtually bound to take place. It is no accident that the dwellings most badly affected by condensation are the various council dwellings which are being erected throughout the country. Because such dwellings have to be subsidised by the rates, it is everyone's

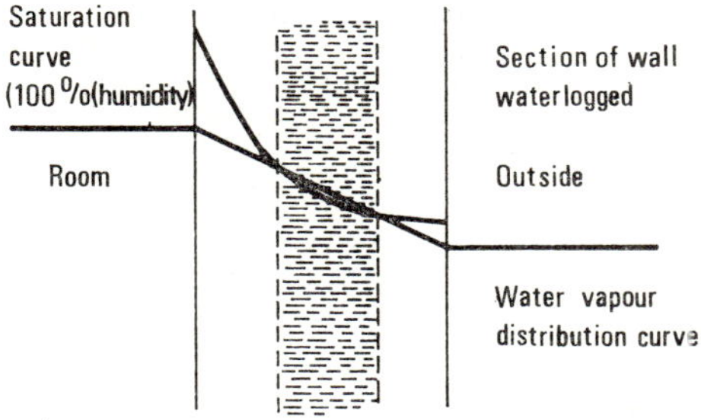


Fig. 25 Unless a vapour barrier is used on the inside of the wall, it is possible for water vapour to condense inside the wall and to saturate the insulation layer

endeavour to throw them up as cheaply as possible, and save a few pennies here and a few pennies there. However, even in this case, it is a matter of being penny wise and pound foolish. Condensation is usually of such a high order that it becomes essential afterwards to carry out expensive remedies and repairs, which are in all cases very much dearer than if the buildings had been designed properly in the first case.

*Calculation of ventilation rate needed*[24]

The minimum number of air changes which is required in a certain room to prevent condensation taking place against the wall is given by the following formula:

$$n = \frac{(W-w)\,(p_i - p_s)}{V\,h\,(p_s - p_o)}$$

where $n$ is the number of room changes of air per hour of ventilation needed

$W$ is the weight of water vapour released per hour in g

$w$ is the weight of water vapour diffusing through walls, ceiling, etc. in g/h

$V$ is the volume of the room in $m^3$

$h$ is the carrying capacity of the air inside the room in $g/m^3$ for water vapour

$p_i$ is the water-vapour pressure inside the room

$p_s$ is the saturation water-vapour pressure at the temperature of the wall surface

$p_o$ is the actual water-vapour pressure of the external air

$p_i$, $p_s$ and $p_o$ are given in $N/m^2$.

All these values and $h$ can be read off from Table 18.

WORKED EXAMPLE

Give the minimum number of air changes required to ventilate a room measuring 4 metres × 3 metres × 3 metres if water vapour totalling 2800 grammes is released per hour. We assume that the loss of water vapour by diffusion processes through walls, ceiling and imperfections of window and door constructions amounts to 350 grammes per hour. The internal air temperature is equal to 20°C and the internal relative humidity is 86%. The external air temperature is 2°C with a relative humidity of 90% and the walls are insulated in such a way that the internal surface temperature of the most exposed wall is equal to 14°C.

$w = 2450$ g $V = 36$ m$^3$
$h = 17{\cdot}9$ g/m$^3$ $\quad\quad p_i = 2330$ N/m$^2 \times 0{\cdot}86 = 2010$ /Nm$^2$
$p_s = 1590$ N/m$^2$ and $\quad\quad p_o = 0{\cdot}9 \times 703 = 634$ N/m$^2$

Let us now substitute in our equation:

$$n = \frac{2450 \times (2010 - 1590)}{36 \times 17{\cdot}9 \times (1590 - 634)} = 1{\cdot}47$$

Therefore, if there is sufficient ventilation capacity to change the air about $1\frac{1}{2}$ times each hour, no condensation should take place against a surface with surface temperature 14°C.

## Designing condensation-free dwellings

Let us once more consider the basic ventilation equation:

$$n = \frac{(W - w)\,(p_i - p_s)}{V\,h\,(p_s - p_o)}$$

where $n$ is the number of room changes of air per hour
$W$ is the hourly rate of water released in the room in question in g
$w$ is the hourly rate of water diffusion from the room in question in g
$V$ is the volume of the room in m$^3$
$h$ is the carrying capacity of the air inside the room in g/m$^3$ of water-vapour
$p_i$ is the water-vapour pressure inside the room
$p_s$ is the saturation water-vapour pressure corresponding to the temperature of the surface
$p_o$ is the actual water-vapour pressure of the external air
$p_i$, $p_s$ and $p_o$ are given in N/m$^2$

To keep the ventilation rate $n$ a minimum, without incurring the risk of condensation, it therefore becomes necessary to:

(1) keep $W - w$ as small as possible
(2) keep $p_i - p_s$ as small as possible
(3) keep $V$ as large as possible
(4) keep $h$ as large as possible
(5) keep $p_s - p_o$ as large as possible.

Interpreting all this qualitatively, we can see that the minimum amount of condensation risk occurs when

(1) The amount of water evaporating in a room is a minimum while porosity of walls is high. This means that one must always aim at well-ventilated external siding and roofing when using such inherently impervious materials such as glass, metal, and so on.
(2) There should be the minimum amount of difference between the surface temperature on the inside of the wall and the room air itself. This is, of course, another way of saying that the wall must be well insulated.
(3) Large rooms suffer less from condensation than small rooms.
(4) The carrying capacity *h* is large when the internal temperature is high. Cold rooms suffer very much more from condensation than warm rooms.
(5) There should be the maximum amount of difference in temperature between the outside air and the internal surface of the wall. In other words, good insulation again.

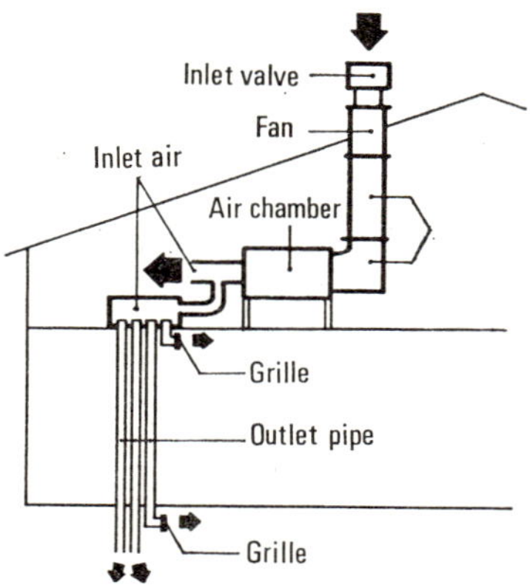


FIG. 26 The Bahco scheme of ventilation gives a sufficient number of air changes, when combined with adequate insulation, to make condensation impossible

All this is, of course, common sense. But this should show that, if mathematical reasoning is right, then it must always back up common-sense reasoning. Mathematical reasoning is superior to common-sense reasoning in that it gives you a quantitative answer.

To evaluate the approximate internal surface temperature we use the following equation:

$$\theta_s = \frac{8{\cdot}5\ \theta_i - U\ (\theta_i - \theta_o)}{8{\cdot}5}$$

where $\theta_s$ is the internal wall surface temperature in °C
$\theta_i$ is the room temperature in °C
$\theta_o$ is the external air temperature in °C
$U$ is the $U$-value of the wall or other section in $W/m^2$ K

WORKED EXAMPLE

Calculate the internal surface temperature of a solid brick wall with $U$-value 2·27 $W/m^2$ K if the external temperature equals 0°C and the room temperature equals 20°C

$$\theta_s = \frac{8{\cdot}5 \times 20 - 2{\cdot}27\ (20 - 0)}{8{\cdot}5} = 14{\cdot}7°C$$

Table 21 gives the internal surface temperatures of a number of windows and walls, which are applicable when there are average wind conditions, and when the external temperature equals 0°C, while the internal room temperature is 20°C. The values quoted for internal wall surface temperatures are the accurate ones.

Interpreting these results we can see that condensation takes place against a single-glazed window whenever the relative humidity of air at 20°C exceeds the very low figure of 43·5%. The reason for this is that the dew point under the conditions mentioned is only 7·2°C. Double glazing raises the dew point to 13·2°C and ensures that no condensation against windows takes place until the relative humidity exceeds 63·6%. Even after condensation conditions have been reached, there is far less run-off of water with better insulated walls and windows than with poorly insulated ones. Let us assume that the room in question had a relative humidity of 75% at a temperature of 20°C, and that the outside temperature is, as before, 0°C.

If the window is single-glazed, the surface temperature on the inside is 7·2°C corresponding to a carrying capacity of 7·95 g of water per $metre^3$. The carrying capacity of 1 $metre^3$ of fully saturated air at 20°C equals 17·9 g of water, and that of 75% relative humidity air is $17{\cdot}9 \times 0{\cdot}75 = 13{\cdot}4$ g. For each $metre^3$ of

TABLE 21. The behaviour of various walls from the condensation point of view

| *Type of wall, etc.* | U-*value of wall* ($W/m^2\ K$) | *Internal surface temperature* (°*C*) | *Maximum % relative humidity permitted in room without condensation taking place* |
|---|---|---|---|
| Single-glazed window | 5·7 | 7·2 | 43·5 |
| Double-glazed window | 2·85 | 13·2 | 63·6 |
| 230 mm solid brick wall | 2·27 | 14·5 | 73·4 |
| Brickwork/cavity/aerated concrete wall | 1·13 | 17·3 | 84·1 |
| 280 mm cavity brick wall with cavity filled using either urea-formaldehyde foam, polyurethane foam or mineral wool, 50 mm thick | 0·57 | 18·5 | 90·9 |
| Brick wall filled with 100 mm of mineral wool | 0·34 | 19·0 | 94·2 |

air considered, the run-off of water against the window pane is equal to $13{\cdot}4 - 7{\cdot}95 = 5{\cdot}45$ g.

We must not forget that when kettles are boiling continuously, there is a complete cycle of air which comes into contact with the windows, has some of its moisture condensed, and then passes above the kettle to pick up more water vapour. Hence the big puddles underneath windows.

If the window is double-glazed, the surface temperature on the inside is 13·2°C, corresponding to a carrying capacity of 11·65 g water vapour/metre$^3$. Assuming the same room conditions as before, the run-off per metre$^3$ of air is now only $13{\cdot}4 - 11{\cdot}65 = 1{\cdot}75$ g. As the circulation of the air past the window in question is still taking place at the same rate, it is obvious that the amount of condensation taking place is much reduced. To start off with, there is no condensation at all until the relative humidity exceeds 63·6%. But even if the relative humidity increases to the high figure of 75% the amount of water running off is only:

$\frac{1{\cdot}75 \times 100}{5{\cdot}45}\% = 32\%$ of the quantity of condensation occurring with single-glazed windows.

The same arguments in favour of better insulation standards can be advanced for walls. For example, no condensation will take place against a brick wall filled with 100 mm of mineral wool as insulation, unless the relative humidity rises to 94·2%, which is roughly the humidity inside a Turkish bath.

*Clothes drying*

One of the major sources of water vapour in dwellings, and in consequence a source of condensation, is the enormous amount of water contained in clothes after they have been washed. The following are experimental figures, obtained by the author, of the weight of water contained in typical items of clothing after washing and careful spin drying. Naturally such figures can only be approximate but may serve as a guide regarding the actual amount of water to be driven off during a typical family laundry.

| | |
|---|---|
| Pair of stocking tights | 15 g |
| Woman's blouse | 60 g |
| Baby's nappy | 100 g |
| Child's dress | 120 g |
| Man's shirt | 160 g |
| Towel | 220 g |
| Man's heavy sweater | 300 g |
| Dressing gown | 480 g |

It is obvious that rooms in which clothes are being dried must have very good ventilation facilities. A solution in the average dwelling may be a bathroom with a mechanical ventilation fan, equipped with clothes line, clothes racks, etc. Drying cabinets must have a vent to the outside, as otherwise the damp air released finds its way back into the dwelling. If the drying cabinet is heated and the vent is at the top, one need not provide mechanical ventilation as the difference in the relative density between the colder inlet air at the bottom of the drying cabinet and the warmer outlet air ensures removal of exhaust air. The outlet air is able to carry a good deal of water vapour because it is much warmer than the normal ambient air temperature in the room. It is essential that a hermetic seal is provided between the drying cabinet and the room, otherwise moisture-laden air can penetrate into the dwelling, with extremely unpleasant results. The inlet vent to the drying cabinet should be positioned as low as possible.

The practice of drying clothes in the kitchen should be discouraged. The kitchen ventilation system would not be able to cope with such huge quantities of extra moisture, in addition to the normal water vapour produced during cooking, washing and washing-up.

*Kitchen ventilation*

Because of the enormous quantities of water vapour which are being evaporated in the average kitchen, it becomes essential to employ extra forced ventilation to extract this moisture, as otherwise it may drift to other rooms, causing condensation against windows or even against walls and ceilings. But a fan mounted simply against the window or at the ceiling is really not particularly effective. One should always use a fume hood in a kitchen via which all the cooking vapours are dissipated directly to the outside. In many European countries a fume hood is specified in the standard building ordinances, yet in Britain it is not usually found even in luxury dwellings. The author considers that a fume hood is an essential item in a kitchen, just as a water flush is an essential item in a toilet. What happens is that the air above the cooker is very hot and carries a large quantitity of water vapour, which is extracted directly to the outside without being distributed in the room.

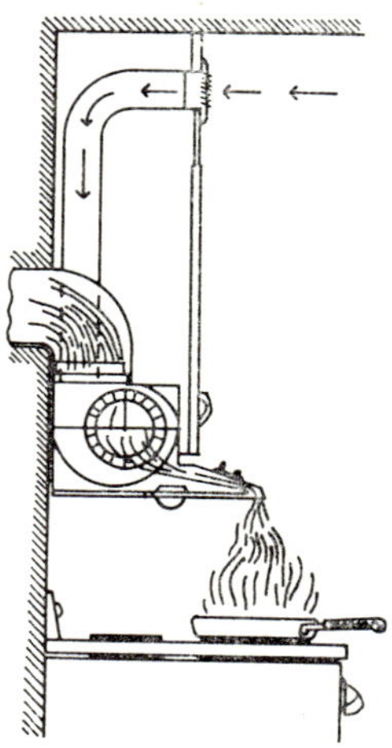

FIG. 27 A ventilation system with a fume hood removes both water vapour and distilled fat straight from the cooker and pumps them outside before they become dispersed in the kitchen

But this is not the only function of a fume hood. The quantities of grease which are being liberated during cooking processes are enormous. It can be calculated, using standard chemical engineering mathematics, that when fat meat is boiled, 6·5 g of fat are distilled for every 1000 g of water evaporated. When food is fried, the figure is several times as high as this. In the absence of a fume hood the fat is distributed virtually throughout the kitchen. When there is a fume hood the fat vapours are led directly to the outside, leaving the kitchen clean.

*Mould formation*

Most moulds thrive on wet conditions, and spread under conditions of condensation against walls, floors and so on. Their food consists of carbohydrates (such as wallpaper) or fats such as deposited grease. To avoid mould formation, it is necessary to avoid condensation.

## Temporary condensation[18]

Condensation can take place against walls of houses either on the inside or the outside if there should be a drastic change in temperature. For example, condensation occurs on the outside wall of a building in the morning, after a very cold night. The reason is that the wall has a very low surface temperature, while the air, which is now quite warm, has a high carrying capacity for moisture. Hence the reason for the sudden condensation against the outside wall. In general such a form of condensation does not last long. As the wall is heated up by the sun, the moisture evaporates. The same can happen inside a dwelling if heaters are suddenly turned on, and the air is at the same time kept fairly humid (cooking, washing, etc. of the inhabitants). If the internal surface of an exterior wall is still cold, condensation may take place against it. Again this condensation evaporates off as the wall heats up, provided, of course, that the insulation of the wall is good enough to retain this heat without being dissipated to the outside by conduction and convection.

## How to cure dampness in dwellings

As with most things, prevention is far better than cure. It is a good deal better and cheaper to design dwellings which are free from condensation than to try afterwards to cure the complaint.

But many houses and flats have now been built which suffer so severely from condensation as to be virtually uninhabitable. It is therefore necessary to examine methods of cure as well.

First of all we must make sure that the troubles experienced are due to condensation and not due to rainwater penetration from the outside, a leaky gutter or roof, or even a burst pipe. The dampness due to a burst pipe can be detected by cutting off the water at the main and seeing whether the dampness experienced reduces markedly overnight.

Rainwater penetration is a maximum after a really heavy rain storm. Condensation, on the other hand, is worst after a period of prolonged cold and dry weather. Also it is generally spread over a far larger area than rainwater penetration due to defective flashing and so on. Common sense should usually be enough to tell the source of dampness damage.

To reduce condensation the following three actions should be taken, either singly or together:

(1) the ambient temperature of the inside of the dwelling should be increased
(2) the air-change rate should be stepped up
(3) the insulation rating of the affected walls, etc. should be increased

It is obvious that one must try the methods which give the biggest improvement at the lowest cost.

*Increasing the ambient temperature*

By increasing the inside temperatures in dwellings one can drastically increase the carrying power for water vapour of the air. A rise in temperature from 12°C to 20°C increases the carrying power by no less than 50% and is often enough to stop condensation under most circumstances. One should consider installing systems of background heating such as water-filled radiators, off-peak electric heating systems, etc. Bedrooms are frequently particularly subject to heavy condensation. They are kept unheated all day, and often also at night, yet large quantities of water vapour are released there every night. Warm bedrooms are not only conducive to the avoidance of illnesses such as rheumatism, lung disorders, and so on; they are also much less likely to suffer from condensation.

Nearly all European building specifications recommend that the ambient temperatures inside a bathroom and toilet should be between 1 and 3°C higher than in a living room. Yet the British Ministry of Housing and Local Government, in its wisdom or otherwise, recommends that a bathroom should be 5°C *colder* than a living room. One need look no further for reasons why bathrooms and toilets are so badly plagued by condensation. The author would recommend instead that the European norms should be followed. This could best be achieved by incorporating a sensibly sized hot-water radiator, which could at the same time act as a towel rail, in such rooms. Failing this, one can use electrically heated towel rails, but great care must be taken in the choice and installation of such equipment, due to the inherent danger of having electricity in 'wet' rooms. Ceiling infra-red heaters are also of value.

*Increasing the air-change rate*

Attempts have frequently been made to cure condensation simply by leaving gaps all over the place, like making it impossible to shut windows, putting in air bricks at odd positions and removing draught proofing round doors. Such methods are often counter-productive. Admittedly the air-change rate is increased, but it is not always the most saturated air which is removed. Instead the increase in draughts often means that the internal temperature falls and that in consequence the carrying power of the air for water vapour reduces. The result—more condensation.

Ventilation should always be practised in such a way that hot moisture-laden air only is removed. The answer would be mechanical ventilation systems in the kitchen and bathroom. Electric fans equipped with shutters are mass produced and in consequence cheap to buy. They can be fitted either to windows, which is a cheaper method but not as good, or against the wall. When the electric fans are combined with a fume hood they are most effective in removing the bulk of the cooking vapours straight from the stove to the outside. In bathrooms extractor fans are also extremely useful in removing the huge quantities of water vapour produced when someone has a hot bath.

*Improving insulation*

This is the most direct way of preventing condensation. It is a commonly-held fallacy that the improvement of insulation of one item of building construction causes the worsening of condensation elsewhere. This is not so. If one stops water dripping down windows, the water is not going to start dripping down the walls instead. Naturally one must start insulating wherever the effect of condensation is worst. This is most frequently on the windows. Large modern picture windows are particularly liable to condensation troubles.

To cure condensation on windows one constructs light frames of the size of each window light. The cross section of the frame is 25 mm × 20 mm and the joints are simple dowel joints. Next 10 mm × 10 mm triangular hardwood dowelling is pinned to the insides of the frame, using a water-resistant glue and panel pins. These frames are then hinged loosely to the *outside* of the existing windows, using hinges which are simply screwed on to the timber and not sunk in. The second frames can be either side hung or top hung. Nor does it matter if the windows are fixed lights or opening lights. Such frames can be installed equally easily on metal windows, except that in such a case holes must be drilled into the metal so that they miss the glass. The holes must be suitably tapped before fixing the hinges on with metal screws. On the side opposite the hinges the frame is secured by one or two hooks and eyelets. The frame which is so fixed has a considerable gap between it and the base frame. This is caused by the thickness of the non-recessed hinges. The gap ensures good ventilation of the cavity between the two panes so that steaming up can never happen, even under the most drastic conditions. Only the cheapest window glass is normally needed for the second frame unless the window is very large, in which case one may have to use thicker glass.

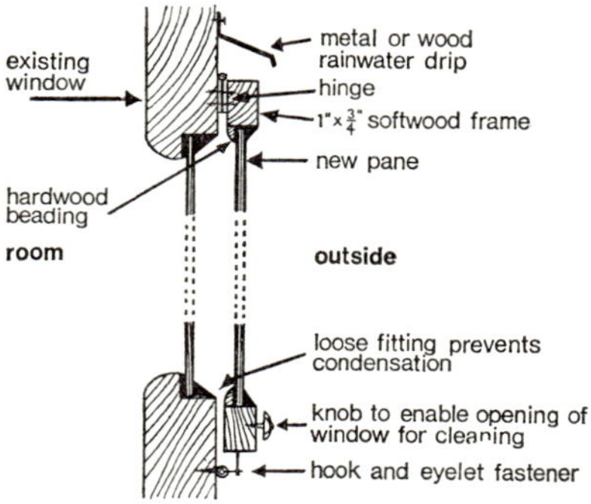


FIG. 28 An extra pane on outside of existing window prevents condensation

As there is a gap between the two frames it is necessary to protect the top edge against direct penetration by rainwater. If the window already has a good overhang, everything is fine. But where the overhang is poor or non-existent, it is necessary to position either a timber or metal strip above the window, so that rainwater runs off clear of the second unit. In most cases this is not a difficult job. To clean the space between the panes, the hooks are opened, the panes cleaned and the hooks shut again. Outward-opening windows operate quite normally, the two window frames swinging as one. A system like this completely eliminates condensation on windows.

Condensation on walls can be eliminated either by filling the cavity of the wall with a high-void material such as ureaformaldehyde foam, polyurethane foam, blown-in mineral wool or polystyrene pellets, if the construction of the wall allows it, or by fixing additional insulation on the inside. Slabs of expanded polystyrene with internal wall surfaces attached can be fixed easily to the existing walls by gluing. The surface which faces the room is either hardboard, fibreboard or plasterboard, and can be decorated directly. The thickness of expanded polystyrene used should not be less than 10 mm.

Condensation on floors is best eliminated by the use of carpets. Alternatively, one can fix battens, place expanded polystyrene or mineral wool into the spaces in between, and cover the top with chipboarding, plywood or even tongued and grooved boarding. This technique is a good one to carry out with many existing houses and flats where the ground floor simply consists of plastic tiles stuck on to a concrete slab, which has no insulation whatever downwards. Condensation on such floors is often very bad, particularly against walls.

# 5 Conclusion

## Improving our internal environmental standards

Housing in Great Britain is of two types:

(1) housing which is built privately either for sale or for letting
(2) municipal housing

*Privately built housing*

The standard of privately built housing has gradually crept up. People are becoming more comfort-conscious and builders have had to improve their standards to be able to sell their products. Central heating at one time was considered a luxury and few houses had such a feature. Nowadays few people would buy a new house without it. Insulation standards have also been improved. Many new houses have double glazing in at least some windows, and cavity insulation is also quickly becoming popular. Better education of the public will soon make them become far more particular regarding the type of dwelling they want to have, and the builder will have to comply or go bankrupt. As the standard of living increases people will wish to have better houses, and new improvements should become features of building, as in any other facet of human endeavour. In the private sector at least, competition and natural market forces should be sufficient to ensure improvement.

*Municipal housing*

The British council-house system is conducive to bad building. As such dwellings are subsidised both by the central government and the local authorities, extremely rigid yardsticks exist to keep the price of such houses below a given figure. At such low-cost figures one must of necessity skimp on something. In most cases one skimps on insulation and ventilation, thereby causing the massive condensation troubles found in such houses. In addition, the lack of insulation practised causes very high fuel bills for tenants who, being lower paid than the average population, cannot really afford such bills. They start saving on fuel, thereby lowering the ambient temperatures in their dwellings, and the result is heavy condensation and damage of internal features. Nor does the council particularly care about the fact that the tenants have to use quite uneconomical quantities of fuel to counteract the atrocious insulation of their dwellings. It does not pay for it. Admittedly, maintenance costs of damp-damaged dwellings are large, but this seems to come out of a different account than the capital acquisition of the dwellings. Again no one seems to care, although both capital costs and running costs are in the end paid for by the same people, the taxpayer and the ratepayer.

The writer of this book is a technical man, and as such perhaps not too versed in political and sociological questions, but he would nevertheless make the following suggestions to improve the situation:

(1) Subsidies on council houses should be according to the income of the tenant and not according to the type of house. Poor people should be given a certain rent allowance, depending upon their degree of need, and council houses should then be let at their economic level, being competitive in price and quality with houses built by private builders. This would mean that council houses would cease to be substandard dwellings for socially inferior persons (or persons considered by the owner-occupier to be socially inferior), but simply dwellings suitable for people who wish to remain mobile and therefore do not desire to purchase their own homes. Each family would then balance its own budget against the quality of house or flat it desires. One family may decide not to run a car but instead live in a really well-insulated and well-constructed home. Another family may decide to take the cheapest house available, and spend the rest of their money on other things. Provided the really poor are catered for by generous rent allowances, this system would be much better than the present one.
(2) The system should be adopted in Great Britain, as it is in

nearly all countries in Europe, that heating costs should be part of the rent. If this system is practised, it immediately means that the letting authority becomes vitally interested in the optimum insulation of its dwellings. In Europe most modern estates are district heated, i.e. hot water for running radiators is supplied centrally to each dwelling by means of underground supply-and-return lines. Each dwelling is fitted with a meter, able to measure the heat supplied to it. The rental charges includes a fairly generous amount for heating, but the tenant is entitled to a rebate if his heat consumption is less than the norm allowed. This encourages him to practise personal restraint in the use of heat, i.e. by keeping windows closed when heaters are on, shutting doors, etc. In practice, nearly every tenant except the most careless will be entitled to some rebate, which is usually paid around Christmas, at a time when the money is most welcome.

(3) Capital accounts and maintenance accounts for dwellings should be merged. It will then become obvious that a house constructed at a low initial outlay may cost more over a certain period of time than another house which is constructed better. Perhaps a government contribution towards the cost of housing could be in the form of an annual allowance per dwelling occupied, instead of an interest allowance on the capital for each house erected.

(4) It has become obvious to the author of this book that many of the British building specifications do not seem to be firmly based on scientific fact. This is in marked contrast to, say, the German DIN specifications, particularly DIN 4701[26] (calculation of thermal demand for buildings) and the Swedish BABS specifications. He would earnestly suggest that full use of these specifications and the massive theory behind them should be made in drawing up future British specifications. He would suggest that if present-day council houses in this country were built to accord with German, Dutch, Swedish or Danish building specifications, condensation and other troubles would be a thing of the past.

# References


1. Diamant, R. M. E., *Understanding SI Metrication*, Angus and Robertson, London (1970).
2. Voss, C. C. W., *A synopsis of Physiology*, John Wright, London (1954).
3. Adamson, B. and Höglund, I., *Tekniska och ekonomiska synpunkter på treglassfönster*, Byggmästaren, B7 (1956), B9 (1957).
4. Höglund, I. and Stephenson, D. G., *Tables for the calculation of Solar Radiation on Buildings*, Byggmästaren 9 (1968).
5. Angus, T. C., *The Control of Indoor Climate*, Pergamon, Oxford (1967).
6. *Which Guide to Central Heating*, Consumers' Association (1970).
7. Wilkes, G. B., *Heat Insulation*, John Wiley, New York (1950).
8. McAdams, W. H., *Heat Transmission*, McGraw-Hill, New York (1954).
9. Nash, G. D. *et al.*, *The Thermal Insulation of Buildings*, Her Majesty's Stationery Office (1955).
10. *Space and Water Heating No. 3, Part 8*, (1967). Ministry of Housing and Local Government.
11. Schüle, W., *Wärmetechnische Anforderungen an Fussboden und Decken* Boden, Wand und Decke 4 (1959).
12. Nevander, L. E., *Köldbryggor i Ytterväggar*, Kungliga Tekniska Högskolan, Stockholm (1961).
13. Technical publications issued by the Aluminium Federation.
14. Diamant, R. M. E., *Thermal and Acoustic Insulation of Buildings*, Iliffe, London (1965).

15. Taylor, G., Paper on Hypothermia read at District Heating Conference, London (1968).
16. Diamant, R. M. E., and McGarry, J., *Space and District Heating*, Iliffe, London (1969).
17. Korsgaard, V., *Varmebesparelser ved isolering*, Ingeniören, Copenhagen (1959).
18. Billington, N. S., *Thermal Properties of Buildings*, Cleaver Hume, London (1952).
19. Olsen, F., *Thermal Insulation of Buildings*, OECD Report (1962).
20. Technical brochure issued by Rockwool Company, (1966).
21. Bayley, T., *A Pocket Book for Chemists*, Spon, London (1948).
22. Levy, F. L., *Insulation*, 21 (1962), 53 (1962), 47 (1963).
23. Kaines, C. C., *Moisture in Roofs*, Her Majesty's Stationery Office (1969).
24. Diamant, R. M. E., *Condensation in Dwellings*, paper read at Cardiff Thermal Insulation Symposium (1968).
25. *Heating, Ventilating and Air Conditioning Guide*, ASHVE (1969).
26. *Regeln für die Berechnung des Wärmebedarfs von Gebäuden*, DIN 4701.

# Appendix

## Conversion factors

| *Imperial or non-SI unit* | *Factor* | *SI unit* |
|---|---|---|
| inch | 25·4 | mm |
| inch$^2$ | 645·16 | mm$^2$ |
| inch$^3$ | 16·387 | cm$^3$ |
| foot | 0·304 8 | m |
| foot$^2$ | 0·092 9 | m$^2$ |
| foot$^3$ | 0·028 317 | m$^3$ |
| yard | 0·914 4 | m |
| yard$^2$ | 0·836 13 | m$^2$ |
| yard$^3$ | 0·764 555 | m$^3$ |
| pint (UK) | 0·568 26 | dm$^3$ (litres) |
| gallon (UK) | 4·546 | dm$^3$ (litres) |
| ounce | 28·35 | g |
| pound | 0·453 59 | kg |
| hundredweight (UK) | 50·802 35 | kg |
| ton (UK) | 1016·047 | kg |
| ton of refrigeration | 3·140 045 | kW |
| degree Fahrenheit | $°C = (°F - 32) \times 5/9$ | |
| | $K = (°F - 32) \times 5/9 + 273·15$ | |
| horsepower | 0·746 | kW |
| millimetre of mercury pressure | 133·32 | N/m$^2$ |
| millimetre water gauge | 9·806 3 | N/m$^2$ |
| inch of mercury pressure | 3·386 | kN/m$^2$ |
| inch of water gauge | 249·082 | N/m$^2$ |
| kg force per metre$^2$ | 9·806 65 | N/m$^2$ |
| pound force/foot$^2$ | 47·880 26 | N/m$^2$ |
| pound force/inch$^2$ | 6·894 76 | kN/m$^2$ |

| *Imperial or non-SI unit* | *Factor* | *SI unit* |
|---|---|---|
| Btu | 1·055 06 | kJ |
| calorie | 4·186 8 | J |
| CHU | 1·899 108 | kJ |
| Btu/hour | 0·293 071 | W |
| Btu/lb | 2·326 | kJ/kg |
| Btu/foot$^2$ | 11·356 | kJ/m$^2$ |
| Btu/pound °F | 4·186 8 | kJ/kg K |
| Btu/foot$^2$ hour °F | 5·678 26 | W/m$^2$ K |
| kcal/m$^2$ hour °C | 1·163 | W/m$^2$ K |
| Btu/foot hour °F | 1·730 07 | W/m K |
| Btu inch/foot$^2$ hour °F | 0·144 228 | W/m K |
| Btu inch/foot$^2$ second °F | 519·22 | W/m K |